Nature's Capacities
and their Measurement

Nature's Capacities and their Measurement

NANCY CARTWRIGHT

CLARENDON PRESS · OXFORD
1989

Oxford University Press, Walton Street, Oxford OX2 6DP
Oxford New York Toronto
Delhi Bombay Calcutta Madras Karachi
Petaling Jaya Singapore Hong Kong Tokyo
Nairobi Dar es Salaam Cape Town
Melbourne Auckland
and associated companies in
Berlin Ibadan

Oxford is a trade mark of Oxford University Press

Published in the United States
by Oxford University Press, New York

British Library Cataloguing in Publication Data
Cartwright, Nancy
Nature's capacities and their measurement
1. Physics. Causation
I. Title
530'.01
ISBN 0-19-824477-0

Library of Congress Cataloging in Publication Data
Cartwright, Nancy.
Nature's capacities and their measurement.
Bibliography.   Includes index.
1. Causality (Physics)   2. Probabilities.
3. Physics—Philosophy.   4. Quantum theory.
5. Econometrics.   I. Title.
QC6.4.C3C37   1989   530'.01   88-33015
ISBN 0-19-824477-0

Set by Colset Private Limited, Singapore
Printed and bound in Great Britain by
Biddles Ltd, Guildford and King's Lynn

To Marian and to Yot Beygh

Acknowledgements

Many of the ideas in this book have evolved in conversation with John Dupré, and the work on abstraction has been deeply influenced by Henry Mendell. Indeed, section 5.5 is taken almost verbatim from a paper which Mendell and I have written together. I have learned about econometrics and its history from Mary Morgan, about exogeneity in econometrics from Anindya Banerjee, and about the classical probabilists and associationist psychology from Lorraine Daston. My concern to embed empiricism in practice has been heightened, and my views have become more developed, by working with the historians of science Norton Wise, Tim Lenoir, and Peter Galison. Part of the research for the book was supported by the National Science Foundation (NSF Grant No. SES–8702931), and it was written while I was a fellow at the Wissenschaftskolleg in Berlin, where Elissa Linke typed it. Corrections throughout are due to J. B. Kennedy, who was helped by Hibi Pendleton. I want to thank the Wissenschaftskolleg, the NSF, and all the other people who have helped.

Contents

Introduction

Science is measurement; capacities can be measured; and science cannot be understood without them. These are the three major theses of this book.

The third thesis could be more simply put: capacities are real. But I do not want to become involved in general issues of realism, instrumentalism, or idealism. Rather, I want to focus on the special case of causes and capacities, and why we need them. They are a part of our scientific image of the world, I maintain, and cannot be eliminated from it. I use the term 'scientific image';[1] but one should not take the narrow view that this image is projected from theory alone. Until recently the philosophy of science has focused primarily on articulated knowledge. To learn what Newtonian physics or Maxwell's electrodynamics teaches, we have turned to what these theories *say*. To answer, 'Are there causal powers in Maxwell's electromagnetic world?' we have studied Maxwell's laws to see if they describe causal powers or not. This is in part what I will do. I claim that the laws of electromagnetic repulsion and attraction, like the law of gravity, and a host of other laws as well, are laws about enduring tendencies or capacities. But it is not all that I am going to do. I arrive at the need for capacities not just by looking at the laws, but also by looking at the methods and uses of science. I maintain, as many do today,[2] that the content of science is found not just in its laws but equally in its practices. We learn what the world is like by seeing what methods work to study it and what kinds of forecasts can predict it; what sorts of laser can be built, or whether the economy can be manipulated. I am going to argue that our typical methodologies and our typical

---

[1] I take the idea of the scientific image from Wilfrid Sellars, who contrasts the world as constructed by science with the world of everyday objects, which he calls the 'manifest' image. Cf. W. Sellars, *Science Perception and Reality* (London: Routledge & Kegan Paul, 1963). This is also the usage of Bas van Fraassen in *The Scientific Image* (Oxford: Clarendon Press, 1983).

[2] Cf. *Science in Context*, 3 (1988); T. Lenoir (ed.), *Practice, Context, and the Dialogue between Theory and Experiment*.

applications, both in the natural sciences and in the social sciences, belong to a world governed by capacities, and indeed cannot be made sense of without them.

My position is opposed to the tradition of Hume. I begin with the assumption that the causal language of science makes sense, and that causality is an objective feature of our scientific image of nature. That does not yet separate me from the Humean tradition. Hume too took causal claims to have an analogue in reality. He began with singular causal claims, looking for some special connection between the individual cause and its effect, a connection that would be strong enough to constitute causation. He failed to find anything more than spatio-temporal contiguity, so he moved to the generic level. This marks the first stage in the Hume programme: (1) for Hume, singular causal facts are true in virtue of generic causal facts. But the programme was far bolder: at the generic level causation was to disappear altogether. It was to be replaced by mere regularity. This is the second thesis of the Hume programme: (2) generic causal facts are reducible to regularities. This book challenges both theses. It begins with the claim that, even if the association is law-like, neither regular association nor functional dependence can do the jobs that causality does. Working science needs some independent notion of causal law.

What kind of a concept could this be? I maintain that the Hume programme has things upside down. One should not start with the notion of generic causation at all. Singular causal claims are primary. This is true in two senses. First, they are a necessary ingredient in the methods we use to establish generic causal claims. Even the methods that test causal laws by looking for regularities will not work unless some singular causal information is filled in first. Second, the regularities themselves play a secondary role in establishing a causal law. They are just evidence—and only one kind of evidence at that—that certain kinds of singular causal fact have happened.[3]

It is the singular fact that matters to the causal law because that is what causal laws are about. The generic causal claims of science are

---

[3] I share this view that probabilities are evidence for causal claims and not constitutive of them with David Papineau, but for quite different reasons. Papineau argues that causal truths must be universal associations and not mere probabilistic ones, whereas I maintain that no regularity of any sort can guarantee a causal claim. See D. Papineau, 'Probabilities and Causes', *Journal of Philosophy*, 82 (1985), 57–74.

not reports of regularities but rather ascriptions of capacities, capacities to make things happen, case by case. 'Aspirins relieve headaches.' This does not say that aspirins always relieve headaches, or always do so if the rest of the world is arranged in a particularly felicitous way, or that they relieve headaches most of the time, or more often than not. Rather it says that aspirins have the capacity to relieve headaches, a relatively enduring and stable capacity that they carry with them from situation to situation; a capacity which may if circumstances are right reveal itself by producing a regularity, but which is just as surely seen in one good single case. The best sign that aspirins can relieve headaches is that on occasion some of them do.

My claims, then, are doubly anti-Humean.[4] I take singular causes to be primary, and I endorse capacities. Nonetheless, I am in sympathy with Hume's staunchly empiricist outlook. I want to argue for the centrality of singular causes and capacities in an empiricist world. That is why this book begins with the sentence 'Science is measurement.' This motto is meant to mark the kind of empiricism I presuppose: a strong practical empiricism, which for better or for worse wants to make a difference to how science carries on. It is a kind of operationalism, but without the theory of meaning. Empiricists have traditionally been concerned with two different questions: (*a*) where do we get our concepts and ideas? and (*b*) how should claims to empirical knowledge be judged? The empiricist answer to the first question is: 'Ideas come immediately from experience.' It is the second question that matters for my project and not the first. Indeed, I shall throughout ignore questions about meanings and the source of ideas. In a sense I will be returning to an early form of British empiricism uncontaminated by the Cartesian doctrine of ideas, an empiricism where causal connections not only made sense but where they were in principle observable. Joseph Glanvill, in his estimable apology for the mechanical philosophy which won him his fellowship of the Royal Society in 1664, tells us that Adam could see these connections distinctly:

Thus the accuracy of his knowledge of natural effects, might probably arise from his sensible perceptions of their causes. . . . His sight could inform him whether the Loadstone doth attract by Atomical Effluviums; . . . The Mysterious influence of the Moon, and its causality on the seas motion, was

---

[4] Though see below, ch. 5.

no question in his Philosophy, no more than a Clocks motion is in ours, where our senses may inform us of its cause.[5]

Like Glanvill and his more scientific colleagues—Robert Boyle or Robert Hooke or Henry Power—I shall not worry about what causal connections are, but ask rather, 'How do we learn about them once our sight has been clouded by the fall from grace?' My concern is not with meaning but with method, and that is why I give prominence to the second empiricist question, where the point is not to ground the concepts of science in pure observation or in direct experience. It is rather to ensure that claims to scientific knowledge are judged against the phenomena themselves. Questions about nature should be settled by nature—not by faith, nor metaphysics, nor mathematics, and not by convention nor convenience either. From Francis Bacon and Joseph Glanvill to Karl Popper and the Vienna Circle empiricists have wanted to police the methods of enquiry to ensure that science will be true to nature. That is the tradition in which I follow.

I look to scientists as well as philosophers for ideas and inspiration. William Thomson (Lord Kelvin) was a physicist who maintained just such an empiricism as I assume here. Kelvin's concepts of work and potential crystallized the shift in structure of late nineteenth-century mechanics, and his views on entropy and waste shaped the newly emerging thermodynamics. He wanted to take over electromagnetic theory too, but he lost in a long battle with Maxwell, a battle fought in part over Maxwell's infamous invention of the unmeasurable and unvisualizable displacement current. Kelvin also laid the Atlantic cable, and that for him was doing science, as much as fashioning theories or measuring in the laboratory. Indeed, the two activities were inseparable for him. The recent biography of Kelvin by Crosbie Smith and Norton Wise describes how practice and theory should join in Kelvin's philosophy. According to Smith and Wise, Kelvin's

deep involvement in the Atlantic cable sheds considerable light on his lifelong rejection of Maxwellian electromagnetic theory. The laying and working of the cable required the complete unity of theory and practice that he had always preached . . . [Kelvin's] natural philosophy . . . opposed the metaphysical to the practical. In proscribing mathematical analogies that extended beyond direct experimental observation, it eliminated the flux-

---

[5] J. Glanvill, *The Vanity of Dogmatizing* (London, 1661), 6–7.

force duality from his flow analogies to electricity and magnetism. Maxwell's theory, by contrast, introduced a physical entity which no one had ever observed, the displacement current. But most specifically, [Kelvin] opposed 'metaphysical' to 'measurable', and it is this aspect that the telegraph especially reinforced. The only aspects of Maxwell's theory that [Kelvin] would ever applaud were those related to measurements.[6]

In this passage we see that Kelvin followed the empiricist convention: what he did not like was called 'metaphysics' and consigned to the flames. Yet the real enemy for Kelvin was not so much metaphysics, as philosophers think of it, but instead a kind of abstract and non-representational mathematics. There is a famous saying, 'Maxwell's theory is Maxwell's equations.' That saying is meant to excuse the fact that Maxwell's theory gives no coherent physical picture. A theory need not do that, indeed perhaps should not. What is needed is a powerful mathematical representation that will work to save the phenomena and to produce very precise predictions. Kelvin called this view about mathematics 'Nihilism'. He wanted the hypotheses of physics to describe what reality was like, and he wanted every one of them to be as accurate and as sure as possible.

It is probably this aspect of his empiricism that needs to be stressed in our contemporary philosophical discourse. Each scientific hypothesis should be able to stand on its own as a description of reality. It is not enough that a scientific theory should save the phenomena; its hypotheses must all be tested, and tested severally. This, then, is an empiricism opposed at once to wholism and to the hypothetico-deductive method. The logic of testing for such an empiricism is probably best modelled by Clark Glymour's bootstrap theory of confirmation:[7] the evidence plus the background assumptions deductively imply the hypothesis under test. But the point is entirely non-technical. Scientific hypotheses should be tested, and the tests should be reliable. They should be powerful enough to give an answer one way or another. The answers will only be as sure as the assumptions that ground the test. But it is crucial that the uncertainty be epistemological and not reside in the test itself. That is why I call this empiricism a kind of operationalism, and stress the idea of

---

[6] C. Smith and N. Wise, *Energy and Empire: A Biographical Study of William Thomson, Lord Kelvin* (Cambridge: Cambridge University Press, 1988), ch. 13.

[7] C. Glymour, *Theory and Evidence* (Princeton, NJ: Princeton University Press, 1980).

measurement.[8] Measuring instruments have this kind of ability to read nature. If the measuring instrument operates by the principles that we think it does, and if it is working properly, and if our reading of its output is right, then we know what we set out to learn. A measurement that cannot tell us a definite result is no measurement at all.

My emphasis on measurement and on the bootstrap methodology should make clear that this empiricism is no kind of foundationalism. It will take a rich background both of individual facts and of general assumptions about nature before one can ever deduce a hypothesis from the data; the thin texture of pure sense experience will never provide sufficient support. Nevertheless, this measurement-based empiricism is a stringent empiricism, too stringent indeed to be appealing nowadays, especially in modern physics.[9] Consider the demand for renormalizable theories in quantum electrodynamics, or the constraints imposed on the range of cosmological models by the desire to eliminate singularities. In both cases it is mathematical considerations that shape the theory, and not judgements imposed by the phenomena: 'nihilism' in Kelvin's language. Einstein is one powerful opponent to views like Kelvin's. Let him speak for the rest:

It is really our whole system of guesses which is to be either proved or disproved by experiment. No one of the assumptions can be isolated for separate testing.[10]

It is not the intention of this book to argue that Einstein or the renormalizers are wrong. I do not want to insist that science must be empiricist. Rather, I want to insist that the practical empiricism of

---

[8] My early papers and lectures on this operationalist-style empiricism used a different terminology and form from that used here. A test of a hypothesis, I maintained, should be *totally reliable*. Davis Baird had taught me to see the exact analogy between tests on one hand and instruments and measurements on the other. Cf. D. Baird, 'Exploratory Factor Analysis, Instruments and the Logic of Discovery', *British Journal for the Philosophy of Science*, 38 (1987), 319–37.

[9] For more about this sort of empiricism and the kind of impact it can have in physics, see N. Cartwright, 'Philosophical Problems of Quantum Theory: The Response of American Physicists', in L. Krüger, G. Gigerenzer, and M. Morgan (eds.), *The Probabilistic Revolution*, ii (Cambridge, Mass., MIT Press, 1987), 417–37.

[10] A. Einstein and L. Infeld, quoted in A. Fine, *The Shaky Game* (Chicago, Ill.: Chicago University Press, 1987), 88–9.

measurement is the most radical empiricism that makes sense in science. And it is an empiricism that has no quarrel with causes and capacities. Causal laws can be tested and causal capacities can be measured as surely—or as unsurely—as anything else that science deals with. Sometimes we measure capacities in a physics laboratory or, as in the gravity probe experiment I will discuss, deep in space in a cryogenic dewar. These are situations in which we can control precisely for the effects of other perturbing factors so that we can see in its visible effects just what the cause can do. But most of the discussion in this book will bear on questions that matter outside physics, in the social sciences, in medicine, in agriculture, and in manufacturing strategies for quality control; that is, in any area where statistics enter. I ask, 'Can probabilities measure causes?' The answer is 'Yes'—but only in a world where capacities and their operations are already taken for granted.

The opening phrase of this introduction was the motto for the Cowles Commission for Economic Research: science is measurement. The Cowles Commission initiated the methods most commonly used in econometrics in America today, and its ideas, in a very primitive form, play a central role in my argument. I will focus on the structures of econometrics in this book, but not because of either the successes or the failures of econometrics as a science; rather because of the philosophic job it can do. We may see intuitively that correlation has something to do with causality. But intuition is not enough. We need an argument to connect probabilities with causes, and we can find one in econometrics.

I have sounded so far as if my principal conflicts were with Hume. In fact that is not true. Unlike Hume, I begin by assuming the current commonplace that science presupposes some notion of necessity: that there is something somewhere in nature that grounds the distinction between a genuine law and a mere accidental generalization. What I deny is that that is enough. Bertrand Russell maintained that science needed only functional laws like the equations of physics and had no place for the notion of cause.[11] I think that science needs not only causes but capacities as well. So I stand more in opposition to Russell than to Hume, or more recently to covering-law theorists like

[11] B. Russell, 'On the Notion of Cause', *Proceedings of the Aristotelian Society*, 13 (1913), 1–26.

C.G. Hempel[12] and Ernest Nagel,[13] who accept laws but reject capacities. The most I will grant the covering-law view is that we need both.

My ultimate position is more radical. The pure empiricist should be no more happy with laws than with capacities, and laws are a poor stopping-point. It is hard to find them in nature and we are always having to make excuses for them: why they have exceptions—big or little; why they only work for models in the head; why it takes an engineer with a special knowledge of real materials and a not too literal mind to apply physics to reality.[14] The point of this book is to argue that we must admit capacities, and my hope is that once we have them we can do away with laws. Capacities will do more for us at a smaller metaphysical price.

The book is organized to take one, step by step, increasingly far from the covering-law view. I begin not with the concept of capacity but with the more innocuous-seeming notion of a causal law. Chapters 1 and 2 argue that causal laws are irreducible to equations and regular associations. Nevertheless, they fit into an empiricist world: they can be measured. Chapter 3 introduces singular causes; and finally Chapter 4, capacities. John Stuart Mill plays a major role in the discussion of Chapter 4 because Mill too was an advocate of capacities, or in his terminology 'tendencies'; and my views and arguments are essentially the same as Mill's in modern guise. Laws about tendencies are arrived at for Mill by a kind of abstraction. That sets the theme for Chapter 5. Abstraction is the key to the construction of scientific theory; and the converse process of concretization, to its application. Covering laws seem irrelevant to either enterprise. Chapter 6 gives a concrete example of a question of current scientific interest where capacities matter: 'Do the Bell inequalities show that causality is incompatible with quantum mechanics?' The question cannot be answered if we rely on probabilities and associations alone. It takes the concept of capacity and related notions of how capacities operate even to formulate the problem correctly.

I close with a word about terminology, and some disclaimers. My

[12] C. Hempel, *Philosophy of Natural Science* (Englewood Cliffs, NJ: Prentice-Hall, 1966).

[13] E. Nagel, *The Structure of Science* (New York: Harcourt, Brace & World, 1961).

[14] See arguments in N. Cartwright, *How the Laws of Physics Lie* (Oxford: Clarendon Press, 1983).

capacities might well be called either 'propensities' or 'powers'.[15] I do not use the first term because it is already associated with doctrines about how the concept of probability should be interpreted; and, although I think that capacities are often probabilistic, I do not think that probability gets its meaning from capacities. I do not use the word 'power' because powers are something that individuals have and I want to focus, not on individuals, but on the abstract relation between capacities and properties. I use the non-technical term 'carries' to refer to this relation: 'Aspirins carry the capacity to relieve headaches' or 'Inversions in populations of molecules carry the capacity to lase'. Does this mean that there are not one but two properties, with the capacity sitting on the shoulder of the property which carries it? Surely not. However, I cannot yet give a positive account of what it does mean—though Chapter 5 is a step in that direction. My aims in this book are necessarily restricted, then: I want to show what capacities do and why we need them. It is to be hoped that the subsequent project of saying what capacities are will be easier once we have a good idea of how they function and how we find out about them. The same should be true of singular causal processes as well; though here the problem is somewhat less pressing, since there are several good accounts already available, notably by Ellery Eells,[16] Wesley Salmon,[17] and the author closest to my own views, Wolfgang Spohn.[18]

[15] For propensities in the sense of dispositions see H. Mellor, *The Matter of Chance* (Cambridge: Cambridge University Press, 1971). For more on powers, see R. Harré and E. Madden, *Causal Powers* (Oxford: Blackwell, 1975).

[16] See E. Eells, *Probabilistic Causality*, forthcoming.

[17] See W. Salmon, *Scientific Explanation and the Causal Structure of the World*, Princeton, NJ: (Princeton University Press, 1984).

[18] See W. Spohn, 'Deterministic and Probabilistic Reasons and Causes', *Erkenntnis*, 19 (1983), 371–96; and 'Direct and Indirect Causes', *Topoi*, forthcoming.

1

How to Get Causes from Probabilities

1.1. Introduction

How do we find out about causes when we cannot do experiments
and we have no theory? Usually we collect statistics. But how do
statistics bear on causality? The purest empiricism, following David
Hume, supposes that general causal truths can be reduced to
probabilistic regularities. Patrick Suppes provides a modern detailed
attempt to provide such a reduction in his probabilistic theory of
causality.[1] Others working on probabilistic causality reject empiricist programmes such as Suppes' altogether. Wesley Salmon is a
good example. For a long time Salmon tried to characterize causation using the concept of statistical relevance. But he eventually
concluded: 'Causal relations are not appropriately analysable in
terms of statistical relevance relations.'[2] Salmon now proposes to use
concepts that have to do with causal processes, like the concepts of
propagation and interaction. What then of statistics? When causes
no longer reduce to probabilities, why do probabilities matter?

This chapter aims to answer that question in one particular
domain—roughly the domain picked out by linear causal modelling
theories and path analysis. It is widely agreed by proponents of
causal modelling techniques that causal relations cannot be analysed
in terms of probabilistic regularities. Nevertheless, statistical
correlations seem to be some kind of indicator of causation. How
good an indicator can they be? The central thesis of this chapter
is that in the context of causal modelling theory, probabilities can
be an *entirely reliable instrument* for finding out about causal
laws. Like all instruments, their reliability depends on a number
of background assumptions; in this particular case, where causal

[1] P. Suppes, *Probabilistic Theory of Causality* (Atlantic Highlands, NJ: Humanities Press, 1970).
[2] W. Salmon, *Scientific Explanation and the Causal Structure of the World*
(Princeton, NJ: Princeton University Press; 1984), 185-6.

conclusions are to be drawn, causal premisses must be supplied. Still, the connection is very strong: given the right kind of background information about other causal facts, certain probabilistic relations are both necessary and sufficient for the truth of new causal facts.

This interpretation of causal modelling theory is controversial. Many social scientists simply assume it and set about using modelling theory to draw causal conclusions from their statistics and to teach others how to do so. Others are more explicit: Herbert Asher in *Causal Modelling* claims: 'Both recursive and non-recursive analysis procedures allow one to conclude that a causal relationship exists, but the conclusion holds only under a restrictive set of conditions.'[3] Another example is the book *Correlation and Causality* by David Kenny. Although at the beginning Kenny puts as a condition on causality that it be 'an active, almost vitalistic, process',[4] this condition plays no role in the remainder of the book, and only two pages later he argues that causal modelling, when properly carried out, can provide causal conclusions:

A third reason for causal modelling is that it can provide a scientific basis for the application of social science theory to social problems. If one knows that X causes Y, then one knows that if X is manipulated by social policy, *ceteris paribus*, Y should then change. However if one only knows that X predicts Y, one has no scientific assurance that when X is changed, Y will change. A predictive relationship may often be useful in social policy, but only a causal relationship can be applied scientifically.[5]

By contrast, a large number of other workers argue that no such thing is possible. Consider, for example, Robert Ling's highly critical review of Kenny's book:

[According to Kenny] the path analyst is supposed to be able to extract causal information from the data that other statisticians such as myself cannot, simply because of the additional causal assumption . . . placed on the model.[6]

Ling maintains that Kenny's attempt to infer causes from statistics has a serious 'logical flaw'.[7] He thinks that the derivation of any

[3] H. Asher, *Causal Modelling* (Beverly Hills, CA: Sage, 1983), 12.

[4] D. Kenny, *Correlation and Causality* (New York: Wiley, 1979), 4.

[5] Ibid. 6.

[6] R. Ling, 'Review of Correlation and Causality', *Journal of the American Statistical Association*, 77 (1982), 489-91.

[7] Ibid. 490.

causal claim is a 'logical fallacy'[8] and concludes: 'I feel obliged to register my strongest protest against the type of malpractice fostered and promoted by the title of this book.'[9]

This chapter will argue that causality is not a malpractice in statistics: given the kinds of very strong assumption that go into causal models, it is possible to extract causal information from statistics. This defence does not, of course, counter the familiar and important criticisms made by many statisticians,[10] that the requisite background assumptions are not met in most situations to which social scientists try to apply causal models; nor does it address questions of how to estimate the 'true' probabilistic relations from available data. In addition there will, not surprisingly, be a number of caveats to add. But the broad import of the conclusions here is that you can indeed use statistics to extract causal information if only the input of causal information has been sufficient.

1.2. Determining Causal Structure

Econometrics is surely the discipline which has paid most attention to how functional laws, probabilities, and causes fit together. For econometrics deals with quantitative functional laws, but it is grounded in a tradition that assumes a causal interpretation for them. Moreover, it is with econometrics that probabilities entered economics. So econometrics is a good starting-place for a study of the relationship between causes and regularities.

I say I begin with econometrics, but in fact I am going to describe only the most primitive structures that econometrics provides. For I want to concentrate on the underlying question 'How do causes and probabilities relate?' To do so, I propose to strip away, as far as possible, all unnecessary refinements and complications. In the end I am going to argue that, at least in principle, probabilities can be used to measure causes. The simple structural models I take from econometrics will help show how. But there is, as always, a gap between principle and practice, and the gap is unfortunately widened by the

---

[8] Ibid. 491.

[9] Ibid.

[10] See e.g. D. A. Freedman, 'As Others See Us: A Case Study in Path Analysis', Technical Report No. 55 (Dept. of Statistics, University of California at Berkeley, 1986). To appear in *Journal of Educational Statistics*.

drastic simplifications I impose at the start. It is not easy to infer from probabilities to causes. A great deal of background information is needed; and more complicated models will require not only fuller information but also information which is more subtle and more varied in kind. So the practical difficulties in using the methods will be greater than may at first appear. Still, I want to stress, these are difficulties that concern how much knowledge we have to begin with, and that is a problem we must face when we undertake any scientific investigation. It is not a problem that is peculiar to questions of causality.

The methods of econometrics from which I will borrow are closely connected with the related tool of path analysis, which is in common use throughout the social and life sciences. I concentrate on econometric models rather than on path analysis for two reasons. The first is that I want to focus on the connection, not between raw data and causes, but between causes and laws, whether the laws be in functional or in probabilistic form. This is easier in econometrics, since economics is a discipline with a theory. Econometrics attempts to quantify the laws of economics, whereas the apparently similar equations associated with path analysis in, for example, sociology tend to serve as mere descriptive summaries of the data. Hence the need to be clear about the connection between causes and laws has been more evident in econometrics.

My second reason for discussing econometrics is related to this: the founders of econometrics worried about this very problem, and I think their ideas are of considerable help in working out an empiricist theory of causality. For they were deeply committed both to measurement and to causes. I turn to econometrics, then, not because of its predictive successes or failures, but because of its comparative philosophical sophistication and self-consciousness. The basic ideas about causal modelling in econometrics will be used in this chapter; and many of the philosophical ideas will play a role in Chapters 3 and 4.

Although econometricians had been using statistical techniques in the 1920s and 1930s, modern econometrics, fully involved with probability, originated at the very end of the 1930s and in the 1940s with the work of Jan Tinbergen, Tjalling Koopmans, and Trygve Haavelmo.[11] The methods that characterize econometrics in the

---

[11] For a good history of econometrics, see M. S. Morgan, *The History of Econometric Ideas* (Cambridge: Cambridge University Press, 1989).

USA, at least until the 1970s, were developed in the immediate post-war years principally at the Cowles Commission for Research in Economics, where both Koopmans and Haavelmo were working. The fundamental ideas of probabilistic econometrics were seriously criticized from the start. The two most famous criticisms came from the rival National Bureau of Economic Research and from John Maynard Keynes. The NBER practised a kind of phenomenological economics, an economics without fundamental laws, so its ideas are not so relevant to understanding how laws and causes are to be connected. To study this question, one needs to concentrate on those who believe that economics is, or can be, an exact science; and among these, almost all—whether they were econometricians or their critics—took the fundamental laws of economics to be causal laws.

Consider Keynes's remarks, near the beginning of his well-known critical review of Tinbergen's work on business cycles:

At any rate, Prof. Tinbergen agrees that the main purpose of his method is to discover, in cases where the economist has correctly analysed beforehand the qualitative character of the causal relations, with what strength each of them operates . . .[12]

Tinbergen does not deny that his method aims to discover the strength with which causes operate. But he does think he can pursue this aim successfully much further than Keynes admits. Given certain constraints on the explanatory variables and on the mathematical forms of the relations, '*certain details of their "influence"* can be given . . . In plain terms: these influences can be measured, allowing for certain margins of uncertainty.'[13] Notice that Tinbergen here uses the same language that I adopt: probabilities, in his view, work like an instrument, to *measure* causal influences. Later in this section I will present a simple three-variable model from Herbert Simon to illustrate how this is done.

Criticisms similar to Keynes's had already been directed against earlier statistical methods in econometrics, notably by Lionel Robbins in his well-known *Essay on the Nature and Significance of Economic Science*. Robbins argued, as Keynes would later, that the causes which operate in economics vary across time:

[12] J. M. Keynes, 'Professor Tinbergen's Method', *Economic Journal*, 49 (1939), 558–68 n. 195.
[13] J. Tinbergen, 'On a Method of Statistical Business-Cycle Research: A Reply, ibid. 50 (1940), 141–54 n. 197. Italics original.

The 'causes' which bring it about that the ultimate valuations prevailing at any moment are what they are, are heterogeneous in nature: there is no ground for supposing that the resultant effects should exhibit significant uniformity over time and space.[14]

Robbins does not here reject causes outright, but he criticizes Tinbergen and the other econometricians on their own ground: the economic situation at any time is fixed by the causes at work in the economy. One may even assume, for the sake of argument, that each cause contributes a determinate influence, stable across time. Nevertheless, these causes may produce no regularities in the observed behaviour of the economy, since they occur in a continually changing and unpredictable mix. A cause may be negligible for a long time, then suddenly take on a large value; or one which has been constant may begin to vary. Yet, despite Robbins's doubts that the enterprise of econometrics can succeed, his picture of its ontology is the same as that of the econometricians themselves: econometrics studies stable causes and fixed influences.

To see how the study proceeds I begin with a simple example familiar to philosophers of science: Herbert Simon's well-known paper 'Spurious Correlation: A Causal Interpretation',[15] which is a distillation of the work of earlier econometricians such as Koopmans. The paper starts from the assumption that causation has a deterministic underpinning: causal connections require functional laws. The particular functional relations that Simon studies are linear. There is a considerable literature discussing how restrictive this linearity assumption is, but I will not go into that, because the chief concern here is clarity about the implications even granted that the linearity condition is satisfied. I will also assume that all variables are time-indexed, and that causal laws are of the form 'X_t causes $Y_{t+\Delta t}$', where $t > 0$. I thus adopt the same view as E. Malinvaud in his classic text *Statistical Methods in Econometrics*. Malinvaud says that cyclical causal relations (like those in Fig. 1.1) arise only

because we have disregarded time lags . . . If we had related the variables to

---

[14] L. Robbins, *The Nature and Significance of Economic Science* (London: Macmillan, 1935). Quoted in T. Lawson, 'Realism and Instrumentalism in the Development of Econometrics', MS, University of Cambridge, Faculty of Economics and Politics, 1987).

[15] H. Simon, 'Spurious Correlation: A Causal Interpretation', in H.M. Blalock (ed.), *Causal Models in the Social Sciences* (Chicago, Ill.: Atherton, 1971), 125.

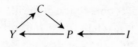

FIG. 1.1

time, the diagram would have taken the form of a directed chain of the following type:

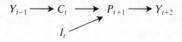

FIG. 1.2

It would now be clear that Y_t is caused by P_{t-1} and causes P_{t+2}.[16]

This assumption plus the assumption that causally related quantities are linear functions of each other generates a 'recursive model', that is, a triangular array of equations, like this:

> ### Structural model
>
> $x_1 = u_1$
> $x_2 = a_{21}x_1 + u_2$
> .
> .
> .
> $x_n = a_{n1}x_1 + a_{n2}x_2 + \ldots + u_n$

Here each x_i is supposed to have a time index less than or equal to that for x_{i+1}, and only factors with earlier indices occur on the right-hand side of any equation. For short, factors on the right-hand side in a structural equation are called the *exogenous* variables in that equation, or sometimes the *independent* variables; those on the left are *dependent*. The *u*s have a separate notation from the *x*s because they are supposed to represent unknown or unobservable factors that may have an effect.

---

[16] E. Malinvaud, *Statistical Methods in Econometrics*, trans. A. Silvey (Amsterdam: North-Holland, 1978), 55.

The model is called 'structural' because it is supposed to represent the true causal structures among the quantities considered. Its equations are intended to be given the most natural causal interpretation: factors on the right are causes; those on the left are effects. This convention of writing causes on the right and effects on the left is not new, nor is it confined to the social sciences; it has long been followed in physics as well. Historian of science Daniel Siegel provides a good example in his discussion of the difference between the way Maxwell wrote Ampère's law in his early work and the way it is written in a modern text. Nowadays the electric current, J, which is treated as the source of the magnetic field, is written on the right—but Maxwell put it on the left. According to Siegel:

The meaning of this can be understood against the background of certain persistent conventions in the writing of equations, one of which has been to put the unknown quantity that is to be calculated—the answer that is to be found—on the left-hand side of the equation, while the known or given quantities, which are to give rise to the answer as a result of the operations upon them, are written on the right-hand side. (This is the case at least in the context of the European languages, written from left to right.) Considered physically, the right- and left-hand sides in most cases represent cause and effect respectively. . .[17]

In the context of a set of structural equations like the ones pictured here, the convention of writing effects on the left and causes on the right amounts to this: an early factor is taken to be a cause of a later, just in case the earlier factor 'genuinely' appears on the right in the equation for the later factor—that is, its coefficient is not zero. So one quantity, x_i, is supposed to be a cause of another, x_j (where $i < j$), just in case $a_{ji} \neq 0$. Grant, for the moment, that for each n, U_n is independent of the other causes of x_n, in all combinations. If so, it will be possible to solve for the values of the parameters a_{ji} in terms of the joint probabilities of the putative causes, and thus to determine which parameters are zero and which are not. But this means that it is possible to infer whether an earlier factor is really a cause of a later factor or not just by looking at the probabilities. This is the point of Simon's paper.

Consider, for example, the three-variable case which Simon discusses:

[17] D. Siegel, 'The Origin of the Displacement Current', *Historical Studies in the Physical Sciences*, 16 (1986), pt. 2, section entitled 'The Solenoidal Current'.

$$x_3 = ax_1 + bx_2 + u_3$$

and imagine that x_2 is correlated with x_3. Does x_2 really cause x_3, or is the correlation spurious? To claim it is spurious is not to say that the correlation itself is not genuine, but rather that the causal relation it suggests is false: x_2 does not in fact cause x_3; the correlation between them occurs because they are both effects of the common cause x_1. According to the convention adopted for reading equations like Simon's, x_2 will be a genuine cause of x_3 just in case $b \neq 0$. (Similarly, x_1 is a genuine cause if $a \neq 0$.) What probabilistic relations must hold for this to be the case?

My discussion will differ from Simon's. His calculations involve joint expectations, for example Exp (x_2x_3), whereas I will solve for b in terms of the expectations *conditional* on fixed values of x_1, like Exp (x_2x_3/x_1). I do so because the expression for the parameters a and b will then be in the form most common in the philosophical literature on probabilistic causality, as well as in the discussion of certain no-hidden-variables proofs in quantum mechanics, which will be described in the last chapter. To solve for b, first note that

$$\text{Exp}(x_2x_3/x_1) = ax_1\text{Exp}(x_2/x_1) + b\text{Exp}(x_2^2/x_1) + \text{Exp}(x_2u_3/x_1)$$

and

$$\text{Exp}(x_3/x_1) = ax_1 + b\text{Exp}(x_2/x_1) + \text{Exp}(u_3/x_1)$$

Using the independence of u_3 from x_1 and x_2 and scaling the error term so that its expectation is zero, this gives[18]

Factorizability

$$b = 0 \text{ iff } \text{Exp}(x_2x_3/x_1) = \text{Exp}(x_2/x_1)\text{Exp}(x_3/x_1)$$

This means that x_2 is not a cause of x_3 just in case their joint expectation factors when x_1 is held fixed. In that case the operation of x_1 as a joint cause for both x_2 and x_3 remains as the only possible account for whatever correlations exist between the two. Hence this criterion is often referred to as the 'common-cause condition'. When more than three variables are involved factorizability continues to mark the absence of a direct influence of one variable on another, only in this case not just x_1 but all the alternative possible causes of the effect-variable must be held fixed. The parameter a can be solved for in a

---

[18] Barring zeros in the denominator.

similar way. It is also easy to generalize to models with more variables. The formulae become more complicated, but as long as the assumptions of the model are satisfied it is always possible to solve for the parameters in terms of the probabilities. It looks, then, as if causes can indeed be inferred from probabilities.

But there is a stock objection: the bulk of Simon's paper is devoted to showing that the parameters can be determined from the probabilities. But the problem occurs one stage earlier, in the interpretation of the data and the selection of variables. The argument given here assumes, roughly, that dependent variables are effects and independent variables are causes. But the facts expressed in a system of simultaneous equations do not fix which variables are dependent and which are independent. Consider for example the equations

$$\begin{array}{ll}
\text{A.} \ x_1 = u_1 & \text{B.} \ x_1 = u_1 \\
\quad x_2 = ax_1 + u_2 & \quad x_2 = ax_1 = u_2 \\
\quad x_3 = bx_1 + u_3 & \quad x_3 = b'x_2 + u_3'
\end{array}$$

These two sets of equations are equivalent assuming $b' = (b/a)$ and $u_3' = u_3 - (b/a)u_2$. Yet they represent different, incompatible causal arrangements. Since Simon seems to argue that causes can be inferred from correlations (when the conditions of the model are met), it is well to look at what he has to say about this problem.

Rather than looking immediately at Simon's work itself, it is instructive to consider a short, non-technical summary of it. The summary is taken from a collected set of student notes from a class on causal modelling taught by Clark Glymour.[19] The notes provide an extremely clear and concise statement of one interpretation of Simon's view and the difficulties it meets. They begin by looking at sets of equivalent equations, like A and B above, which seem to yield different causal pictures. This raises a serious problem for any attempt to ground causal claims purely in functional relationships: 'If altering the equations to a mathematically equivalent set alters the causal relations expressed, then those relations involve additional structure not entirely caught by the equations alone.'[20]

The equations from the Glymour notes are:

$$\begin{array}{l}
\text{IV.} \ a_{11}y_1 + a_{12}y_2 = a_{10} \\
\quad a_{21}y_1 + a_{22}y_y = a_{20}
\end{array}$$

[19] C. Glymour, class notes from a seminar at the University of Pittsburgh, dated 13 Feb. 1983.
[20] Ibid. 4.

$$V. \quad b_{11}y_1 = b_{10}$$
$$a_{21}y_1 + a_{22}y_2 = a_{20}$$

$$VI. \quad b_{11}y_1 = b_{10}$$
$$a_{22}y_2 = c_{20}$$

In these equations:

$$b_{11} = a_{11} - (a_{12}/a_{22})a_{21}$$
$$b_{10} = a_{10} - (a_{12}/a_{22})a_{20}$$
$$c_{20} = a_{20} - (a_{21}/b_{11})b_{10}$$

These, unlike the previous equations, leave no space for omitted factors or random errors—there are no *us*—hence they describe a fully deterministic system. Still, they are intended to be read in the same way: in IV, the variables y_1 and y_2 are causally unordered; in V, y_1 causes y_2; and in VI, the two are causally independent.

The notes continue:

Simon recognizes this problem and offers a solution. He says that if an alteration is made in one of the coefficients of an equation in a linear structure and there exists a variable in that equation whose value is unaltered by the variation, then that variable is exogenous with respect to the other variables in the system. In system IV, 'wiggling' any of the coefficients in the first equation produces a change in both y_1 and y_2. The same is true of the second equation. This system can therefore provide no information about which variable is exogenous. However, in the equivalent system V, if a_{21}, a_{22}, or a_{20} is varied, the value of y_2 will be altered, but y_1 will not be; therefore, y_1 is the exogenous variable under Simon's criterion. This means that y_1 causes y_2. System VI has only one variable in each equation; hence it cannot provide us causal information by this criterion. Simon's contention is that the equivalent system that provides causal ordering in this way is the one that identifies the actual causal relations.

To see that this additional criterion fails to resolve the difficulty, consider the following system which is also equivalent to the three above:

$$VIII. \quad y_2 = (a_{10} - a_{20})/(a_{12} - a_{22}(a_{11}/a_{12}))$$
$$a_{11}y_1 + a_{12}y_2 = a_{10}$$

Using Simon's new criterion, these equations identify y_2 as the independent causal variable because varying any of the coefficients in the second equation produces a change in y_1 but not y_2. Using this sort of rearrangement, it is possible to find equivalent systems for any system of equations in which the causal relations identified by Simon's criteria are completely rearranged.[21]

---

[21] Ibid. 8–9.

So this strategy fails. I do not think that Simon's 'wiggling' criterion was ever designed to solve the equivalence problem. He offered it rather as a quasi-operational way of explaining what he meant by 'causal order'. But whether it was Simon's intent or not, there is a prima-facie plausibility to the hope that it will solve the equivalence problem, and it is important to register clearly that it cannot do so.

Glymour himself concludes from this that causal models are hypo-thetico-deductive.[22] For him, a causal model has two parts: a set of equations and a directed graph. The directed graph is a device for laying out pictorially what is hypothesized to cause what. In recursive models it serves the same function as adopting the convention of writing the equations with causes as independent variables and their effects as dependent. This method is hypothetico-deductive because the model implies statistical consequences which can then serve as checks on its hypotheses. But no amount of statistical information will imply the hypotheses of the model. A number of other philosophers seem to agree. This is, for instance, one of the claims of Gurol Irzik and Eric Meyer in a recent review of path analysis in the journal *Philosophy of Science*: 'For the project of making causal inferences from statistics, the situation seems to be hopeless: almost anything . . . goes.'[23]

Glymour's willingness to accept this construal is more surprising, because he himself maintains that the hypothetico-deductive method is a poor way to choose theories. I propose instead to cast the relations between causes and statistics into Glymour's own bootstrap model: causal relations can be deduced from probabilistic ones—given the right background assumptions.[24] But the background assumptions themselves will involve concepts at least as rich as the concept of causation itself. This means that the deduction does not provide a source for reductive analyses of causes in terms of probabilities.

The proposal begins with the trivial observation that the two sets of equations, A and B, cannot both satisfy the assumptions of the model. The reason lies in the error terms, which for each equation

[22] C. Glymour, R. Scheines, P. Spirtes, and K. Kelly, *Discovering Causal Structure* (New York: Academic Press, 1987).

[23] G. Irzik and E. Meyer, 'Causal Modelling: New Directions for Statistical Explanation', *Philosophy of Science*, 54 (1987), 495–514.

[24] C. Glymour, *Theory and Evidence* (Princeton, NJ: Princeton University Press, 1980).

are supposed to be uncorrelated with the independent variables in that equation. This relationship will not usually be preserved when one set of equations is transformed into another: if the error terms are uncorrelated in one set of equations, they will not in general be uncorrelated in any other equivalent set. Hence the causal arrangement implied by a model satisfying all the proposed constraints is usually unique.

But what is the rationale for requiring the error terms to be uncorrelated? In fact, this constraint serves a number of different purposes, and that is a frequent source of confusion in trying to understand the basic ideas taken from modelling theory. It is important to distinguish three questions: (1) The Humean problem: under what conditions do the parameters in a set of linear equations determine causal connections? (2) The problem of identification (the name for this problem comes from the econometric literature): under what conditions are the parameters completely determined by probabilities? (3) The problem of estimation: lacking knowledge of the true probabilities, under what conditions can the parameters be reliably estimated from statistics observed in the data? Assumptions about lack of correlation among the errors play crucial roles in all three. With regard to the estimation problem, standard theorems in statistics show that, for simple linear structures of the kind considered here, if the error terms are independent of the exogenous variables in each equation, then the method of least squares provides the *b*est *l*inear, *u*nbiased *e*stimates (that is, the method of least squares is BLUE). This in a sense is a practical problem, though one of critical importance. Philosophical questions about the connection between causes and probabilities involve more centrally the first two problems, and these will be the focus of attention in the next sections.

In Simon's derivation, it is apparent that the assumption that the errors are uncorrelated guarantees that the parameters can be entirely identified from the probabilities, and in addition, since any transformations must preserve the independence of the errors, that the parameters so identified are unique. But this fact connects probabilities with causes only if it can also be assumed that the very same conditions that guarantee identifiability also guarantee that the equations yield the correct causal structure. In fact this is the case, given some common assumptions about how regularities arise. When these assumptions are met, the conditions for identifiability and certain conditions that solve the Hume problem are the same. (Probably that is the reason why the two problems have not been clearly distinguished in the econometrics literature.) If they are the same, then causal structure is not merely hypothetico-deductive, as Glymour and others claim. Rather, causes can indeed be bootstrapped from probabilities. This is the thesis of the next two sections.

But first it is important to see exactly what such a claim amounts to in the context of Simon's derivation. Simon shows how to connect probabilities with parameters in linear equations, where the error terms are uncorrelated with the exogenous variables in each equation. But what has that to do with causation? Specifically, why should one think that the independent variables are causes of the dependent variables so long as the errors satisfy the no-correlation assumptions? One immediate answer invokes Reichenbach's principle of the common cause: if two variables are correlated and neither is a cause of the other, they must share a common cause. If the independent variables and the error term were correlated, that would mean that the model was missing some essential variables, common causes which could account for the correlation, and this omission might affect the causal structure in significant ways.

But this answer will not do. This is not because Reichenbach's principle fails to be true in the situations to which modelling theory applies. It will be argued later that something like Reichenbach's principle must be presumed if there is to be any hope of connecting causes with probabilities. The problem is that the suggestion makes the solution to the original problem circular. The starting question of this enquiry was: 'Why think that probabilities bear on causes?' Correlations are widely used as measures of causation; but what justifies this? The work of Simon and others looks as if it can answer this question, first by reducing causation to functional dependence, then by showing that getting the right kinds of functional dependency guarantees the right kinds of correlation. But this programme won't work if just that connection is assumed in 'reducing' causes to functional dependencies in the first place.

There is, however, another argument that uses the equations of linear modelling theory to show why causes can be inferred from probabilities, and the error terms play a crucial role in that argument. The next section will present this argument for the deterministic case, where the 'error' terms stand for real empirical quantities;

cases where the error terms are used as a way to represent genuinely indeterministic situations must wait until Chapter 3.

1.3. Inus Conditions

To understand the causal role that error terms play, it is a help to go back to some philosophically more familiar territory: J. L. Mackie's discussion of inus conditions. An inus condition is an *i*nsufficient but *n*on-redundant part of an *u*nnecessary but *s*ufficient condition.[25] The concept is central to a regularity theory of causation, of the kind that Mackie attributes to John Stuart Mill. (I do not agree that Mill has a pure regularity theory: see ch. 4.)

The distinction between regularities that obtain by accident and those that are fixed by natural law is a puzzling one, and was for Mill as well, but it is not a matter of concern here. Assume that the regularities in question are law-like. More immediately relevant is the assumption that the regularities that ground causation are deterministic: a *complete* cause is sufficient for the occurrence of an effect of the kind in question. But in practical matters, however, one usually focuses on some part of the complete cause; this part is an inus condition. Although (given the assumption of determinism) the occurrence of a complete cause is sufficient for the effect, it is seldom necessary. Mill says: 'There are often several independent modes in which the same phenomenon would have originated.'[26] So in general there will be a plurality of causes—hence Mackie's advice that we should focus on causes which are parts of an *unnecessary* but sufficient condition.

For Mill, then, in Mackie's rendition, causation requires regularities of the following form:

$$E \equiv A_1 X_1 \lor A_2 X_2 \lor \ldots \lor A_n X_n$$

X_1 is a partial cause in Mill's sense, or an inus cause, just in case it is genuinely non-redundant. The notation is chosen with Xs as the salient factors and As as the helping factors to highlight the analogy

[25] J. L. Mackie, *Cement of the Universe* (Oxford: Clarendon Press, 1980), 62.
[26] J. S. Mill, *A System of Logic* (1872), repr. in *Collected Works* (Toronto: Toronto University Press, 1967), vols. VII–VIII, bk. III, ch. x, s. 1.

with the linear equations of the last secton.[27] Mill calls the entire disjunction on the right the 'full cause'. Following that terminology, I will call any set of inus conditions formed from sufficient conditions which are jointly necessary for the effect, a *full* set.

Mackie has a nice counter-example to show why inus causes are not always genuine causes. The example is just the deterministic analogue of the problem of spurious correlation: assume that X_2 and X_3 are joint effects of a common cause, X_1. In this case X_2 will turn out to be an inus condition for X_3, and hence mistakenly get counted as a genuine cause of X_3. The causal structure of Mackie's example is given in Fig. 1.3. Two conventions are adopted in this figure which will be followed throughout this book. First, time increases as one reads down the causal graph, so that in Fig. 1.3, for example, W, A, X_1, B, and V are all supposed to precede X_2 and X_3; and second, all unconnected top nodes—here again W, A, X_1, B, and V—are statistically independent of each other in all combinations. The example itself is this:

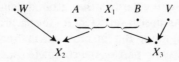

FIG. 1.3

The sounding of the Manchester factory hooters [X_2 in the diagram], plus the absence of whatever conditions would make them sound when it wasn't five o'clock [W], plus the presence of whatever conditions are, along with its being five o'clock, jointly sufficient for the Londoners to stop work a moment later [B], including, say, automatic devices for setting off the London hooters at five o'clock, is a conjunction of features which is unconditionally followed by the Londoners stopping work [X_3]. In this conjunction the sounding of the Manchester hooters is an essential element, for it alone, in this conjunction, ensures that it should *be* five o'clock. Yet it would be most implausible to say that this conjunction causes the stopping of work in London.[28]

---

[27] Notice that this notation is exactly the reverse of that used by Mackie. Also notice that reference to Mackie's background causal field has been dropped, since issues about the field will not play any role here.

[28] Mackie, *Cement of the Universe*, p. 84.

Structurally, the true causal situation is supposed to be represented thus:

$$X_2 \equiv AX_1 \vee W \tag{1}$$
$$X_3 \equiv BX_1 \vee V \tag{2}$$

where the variable V is a disjunction of all the other factors that might be sufficient to bring Londoners out of work when it is not five o'clock. But since $BX_2 \neg W \equiv BAX_1 \neg W$, these two propositions are equivalent to

$$X_2 \equiv AX_1 \vee W \tag{1'}$$
$$X_3 \equiv BX_2 \neg W \vee BX_1 \neg A W \vee BX_1 A W \vee BX_1 \neg A \neg W \vee V. \tag{2'}$$

So X_2 becomes an inus condition. But notice that it does so only at the cost of making $\neg W$ one as well, and that can be of help. If our background knowledge already precludes W and its negation as a cause of X_3, then (2') cannot represent the true causal situation. Though X_2 is an inus condition, it will have no claim to be called a cause.

What kind of factor might W represent in Mackie's example? Presumably the hooters that Mackie had in mind were steam hooters, and these are manually triggered. If the timekeeper should get fed up with her job and pull the hooter to call everyone out at midday, that would not be followed by a massive work stoppage in London. So this is just the kind of thing that $\neg W$ is meant to preclude; and it may well be just the kind of thing that we have good reason to believe cannot independently cause the Londoners to stop work at five every day. We may be virtually certain that the timekeeper, despite her likely desire to get Londoners off work too, does not have access to any independent means for doing so. Possibly she may be able to affect the workers in London by pulling the Manchester hooters—that is just the connection in question; but that may be the only way open to her, given our background knowledge of her situation and the remainder of our beliefs about what kinds of thing can get people in London to stop work. If that were indeed our epistemological situation, then it is clear that (2') cannot present a genuine full cause, since that would make W or its negation a genuine partial cause, contrary to what we take to be the case. Although this formula shows that the Manchester hooters are indeed an inus condition, it cannot give any grounds for taking them to be a genuine cause.

This simple observation about the timekeeper can be exploited more generally. Imagine that we are presented with a full set of inus conditions. Shall we count them as causes? The first thing to note is that we should not do so if any member of the set, any individual inus condition, can be ruled out as a possible cause. Some of the other members of the set may well be genuine causes, but being a member of that particular full set does not give any grounds for thinking so. We may be able to say a lot more, depending on what kind of background knowledge we have about the members of the set themselves, and their possible causes. We need not know all, or even many, of the possible causes. Just this will do: assume that every member of the set is, for us, a possible cause of the outcome in question; and further, that for every member of the set we know that among its genuine causes there is at least one that cannot be a cause of the outcome of interest. In that case the full set must be genuine. For any attempt to derive a spurious formula like (2′) from genuinely causal formulas like (1) and (2) will inevitably introduce some factor known not to be genuine.

Exactly the same kind of reasoning works for the linear equations of causal modelling theory. Each equation in the linear model contains a variable, u, which is peculiar to that equation. It appears in that equation and in no others—notably not in the last equation for the effect in question, which we can call x_e. If these us are indeed 'error' terms which represent factors we know nothing about, they will be of no help. But if, as with the timekeeper, they represent factors which we know could not independently cause x_e, that will make the crucial difference. In that case there will be no way to introduce an earlier x, that does not really belong, into the equation for x_e without also introducing a u, and hence producing an equation which we *know* to be spurious. But if a spurious equation could only come about by transformation from earlier equations which represent true causal connections, the equation for x_e cannot be spurious. The next section will state this result more formally and sketch how it can be proved. But it should be intuitively clear already just by looking at the derivation of (2′) from (1) and (2), why it is true.

More important to consider first is a metaphysical view that the argument presupposes—a view akin to Reichenbach's principle of the common cause. Reichenbach maintained that where two factors are probabilistically correlated and neither causes the other, the two must share a joint cause.[29] Clearly this is too narrow a statement, even for the view that Reichenbach must have intended. There are a variety of other causal stories that could equally account for the correlation. For example, the cause of one factor may be associated with the absence of a preventative of the other, a correlation which itself may have some more complicated account than the operation of a simple joint cause. More plausibly Reichenbach's principle should read: every correlation has a causal explanation.

One may well want to challenge this metaphysical view. But something like it must be presupposed by any probabilistic theory of causality. If correlations can be dictated by the laws of nature independently of the causal processes that obtain, probabilities will give no clue to causes. Similarly, the use of inus conditions as a guide to causality presupposes some deterministic version of Reichenbach's principle. The simplest view would be this: if a factor is an inus condition and yet it is not a genuine cause, there must be some further causal story that accounts for why it is an inus condition. The argument above uses just this idea in a more concrete form: any formula which gives a full set of inus conditions, not all of which are genuine causes, must be derivable from formulae which do represent only genuine causes.

1.4. Causes and Probabilities in Linear Models

Return now to the equations of causal modelling theories.

$$x_1 = u_1$$
$$x_2 = a_{21}x_1 + u_2$$
$$\ldots$$
$$x_n = a_{n1}x_1 + \ldots + u_n$$

One of the variables will be the effect of interest. Call it x_e and consider the eth equation:

$$x_e = a_{e1}x_1 + a_{a2}x_2 + \ldots a_{ee-1}x_{e-1}.$$

In traditional treatments, where the equations are supposed to represent fully deterministic functional relations, the us are differentiated from the xs on epistemological grounds. The xs are supposed

---

[29] H. Reichenbach, *Direction of Time* (Berkeley, Calif.: University of California Press, 1956), 157.

to represent factors which are recognized by the theory; us are factors which are unknown. But the topic here is metaphysics, not epistemology; the question to be answered is, 'What, in nature, is the connection between causal laws and laws of association?' So for notational convenience the distinction between us and xs has been dropped on the right-hand side of the equation for x_e. It stands in the remaining equations, though not to make an epistemological distinction, but rather to mark that one variable is new to each equation: u_i appears non-trivially in the ith equation, and nowhere else. This fact is crucial, for it guarantees that the equations can be identified from observational data. It also provides just the kind of link between causality and functional dependence that we are looking for.

Consider identifiability first. In the completely deterministic case under consideration here, where the us represent real properties, probabilities never need to enter. The relevant observational facts are about individuals and what values they take for the variables x_1 . . ., x_e. Each individual, i, will provide a set of values $\{x_1^i, \ldots, x_e^i\}$. Can the parameters $a_{e1}, \ldots, a_{ee} - _1$, be determined from data like that? The answer is familiar from elementary algebra. It will take observations on $e - 1$ individuals, but that is enough to identify all $e - 1$ parameters, so long as the variables themselves are not linearly dependent on each other.

A simple three-dimensional example will illustrate.

$$x_3 = ax_1 + bx_2$$

Given the values $\{o_1, o_2, o_3\}$ for the first observation of x_1, x_2, and x_3, and $\{p_1, p_2, p_3\}$ for the second, two equations result

$$o_3 = ao_1 + bo_2$$
$$p_3 = ap_1 + bo_2.$$

Solving for a and b gives

$$a = \frac{o_2p_3 - p_2o_3}{p_1o_2 - p_2o_1}$$

$$b = \frac{p_1o_3 - o_1p_3}{p_1o_2 - p_2o_1}$$

Imagine, though, that x_1 and x_2 are linearly dependent of each other, i.e. $x_1 = \Gamma x_2$ for some Γ. Then $p_1 = \Gamma p_2$ and $o_1 = \Gamma o_2$, which implies that $p_1o_2 = p_2o_1$. So the denominators in the equations for a and b are zero, and the two parameters cannot be identified. Because x_1 and x_2 are multiples of each other, there is no unique way to partition the linear dependence of x_3 on x_1 and x_2 into contributions from x_1 and x_2 separately; and hence no sense is to be made of the question whether the coefficient of x_1 or of x_2 is 'really' zero or not. This result is in general true. All $e - 1$ parameters in the equation for x_e can be identified from $e - 1$ observations so long as none of the variables is a linear combination of the others. If some of the variables are linearly related, no amount of observation will determine all the parameters.

These elementary facts about identifiability are familiar truths in econometrics and in modelling theory. What is not clearly agreed on is the connection with causality, and the remarkable fact that the very same condition that ensures the identifiability of the parameters in an equation like that for x_e also ensures that the equation can be given its natural causal reading—so long as a generalized version of Reichenbach's principle of the common cause can be assumed. The argument is exactly analogous with that for inus causes. If the factors with non-zero coefficients on the right-hand side of the equation for x_e are all true causes of x_e, everything is fine. But if not, how does that equation come to be true? Reichenbach's principle, when generalized, requires that every true functional relationship have a causal account. In the context of linear modelling theory, this means that any equation that does not itself represent true causal relations must be derivable from equations that do. So if the equation contains spurious causes, it must be possible to derive it from a set of true causal equations, recursive in form, like those at the beginning of this section. But if each of the equations of the triangular array contains a u that appears in no other equation, including the spurious equation for x_e, this will clearly be impossible. The presence of the us in the preceding equations guarantees that the equation for x_e must give the genuine causes. Otherwise it could not be true at all.

This argument can be more carefully put as a small theorem. The theorem depends on the assumption that each of the xs on the right-hand side of an equation like that for x_e satisfies something that I will call the 'open back path requirement'. For convenience of expression, call any set of variables that appear on the right-hand side of a linear equation for x_e a 'full set of inus conditions' for x_e, by analogy with Mackie's formulae. Only full sets where every member *could* be a cause of x_e, so far as we know, need be considered. Recall that the variables under discussion are supposed to be time-indexed, so that this requires at least that all members of the full set represent quantities that obtain before x_e. But it also requires that there be nothing else in our background assumptions to rule out any of the inus conditions as a true cause.

The open back path assumption is analogous with the assumption that was made about the Manchester timekeeper: every inus condition in the set must have at least one cause that is known not to be a cause of x_e, and each of these causes in turn must have at least one cause that is also known not to be a cause of x_e, and so forth. The second part of the condition, on causes of causes, was not mentioned in the discussion of Mackie's example, but it is required there as well. It is called the 'open back path condition' (OBP) because it requires that at any time-slice there should be a node for each inus factor which begins a path ending in that factor, and from which no descending path can be traced to any other factor which either causes x_e, or might cause x_e, for all we know. This means not only that no path should be traceable to genuine causes of x_e, but also that none should be traceable to any of the other members of the full set under consideration. The condition corresponds to the idea expressed earlier that every inus condition x should have at least one causal history which is known not to be able to cause x_e independently, but could do so, if at all, only by causing x itself.

Fig. 1.4 gives one illustration of this general structure. The inus conditions under consideration are labelled by xs; true causes by ys;

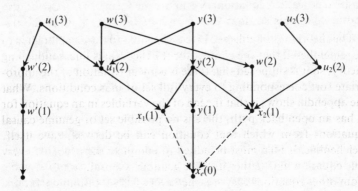

Fig. 1.4

factors along back causal paths by *u*s; and other factors by *w*s. Time indices follow variables in parentheses. In the diagram $u_1(3)$. $u_1(2)$ forms an open back path for $x_1(1)$; and $u_2(3)$. $u_2(2)$, for $x_2(1)$. The path $y(3)$. $u_1(2)$ will not serve as an open back path for $x_1(1)$, since $y(3)$ is a genuine cause, via $y(2)$, of $x_e(0)$, and hence cannot be known not to be. Nor will $w(3)$. $u_1(2)$ do either, since $w(2)$ causes $x_2(0)$, and hence may well, for all we know, have an independent route by which to cause $x_e(0)$.

Notice that these constructions assume the transitivity of causality: if $y(2)$ causes $y(1)$ which causes $x_e(0)$, then $y(2)$ causes $x_e(0)$ as well. This seems a reasonable assumption in cases like these where durations are arbitrary and any cause may be seen either as an immediate or as a more distant cause, mediated by others, depending on the size of the time-chunks. The transitivity assumption is also required in the proof of the theorem, where it is used in two ways: first in the assumption that any x which causes a y is indeed a genuine cause itself, albeit a distant one by the time-chunking assumed; and second, to argue, as above, that factors like $w(2)$ that cause other factors which might, so far as we know, cause $x_e(0)$ must themselves be counted as epistemically possible causes of $x_e(0)$.

The theorem which connects inus conditions with causes is this:

If each of the members of a full set of inus conditions for x_e may (epistemically) be a genuine cause of x_e, and if each has an open back path with respect to x_e, all the members of the set are genuine causes of x_e

where

OBP: $x(t)$ has an *open back path* with respect to $x_e(0)$ just in case at any earlier time t', there is some cause, $u(t')$, of $x(t)$, and it is both true, and known to be true, that $u(t')$ can cause $x_e(0)$ only by causing $x(t)$.

The theorem is established in the appendix to this chapter. It is to be remembered that the context for the theorem is linear modelling theory, so it is supposed that there is a linear equation of the appropriate sort corresponding to every full set of inus conditions. What the appendix shows is that if each of the variables in an equation for x_e has an open back path, there is no possible set of genuine causal equations from which that equation can be derived, save itself. Reichenbach's idea must be added to complete the proof. If every true equation must either itself be genuinely causal, or be derivable from other equations that are, then every full set of inus conditions with open back paths will be a set of genuine causes.

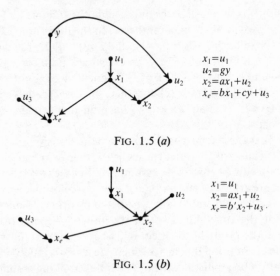

FIG. 1.5 (*a*)

$x_1 = u_1$
$u_2 = gy$
$x_2 = ax_1 + u_2$
$x_e = bx_1 + cy + u_3$

$x_1 = u_1$
$x_2 = ax_1 + u_2$
$x_e = b'x_2 + u_3$.

FIG. 1.5 (*b*)

It may seem that an argument like that in the appendix is unnecessary, and that it is already apparent from the structure of the equations that they cannot be transformed into an equation for x_e without introducing an unwanted u. But the equations that begin this section are not the only ones available. They lay out only the causes for the possibly spurious variables $x_1, x_2, ..., x_{e-1}$. Causal equations for the genuine causes—call them $y_1, ..., y_m$—will surely play a role too, as well as equations for other variables that ultimately depend on the xs and ys, or vice versa. All of these equations can indeed be put into a triangular array, but only the equations for the xs will explicitly include us. A simple example will illustrate. Fig. 1.5(*a*) gives the true causal structure, reflected in the accompanying equations; Fig. 1.5(*b*) gives an erroneous causal structure, read from its accompanying equations, which can be easily derived from the 'true' equations of Fig. 1.5(*a*). So it seems that we need the slightly more cumbersome argument after all.

1.5. Conclusion

This chapter has tried to show that, contrary to the views of a number of pessimistic statisticians and philosophers, you can get from probabilities to causes after all. Not always, not even usually—but in just the right circumstances and with just the right kind of starting information, it is in principle possible. The arguments of this chapter were meant to establish that the inference from probability to cause can have the requisite bootstrap structure: given the background information, the desired causal conclusions will follow deductively from the probabilities. In the language of the Introduction, probabilities can *measure* causes; they can serve as a perfectly reliable instrument.

My aim in this argument has been to make causes acceptable to those who demand a stringent empiricism in the practice of science. Yet I should be more candid. A measurement should take you from data to conclusions, but probabilities are no kind of data. Finite frequencies in real populations are data; probabilities are rather a kind of ideal or theoretical representation. One might try to defend probabilities as a good empirical starting-point by remarking that there are no such things as raw data. The input for any scientific inference must always come interpreted in some way or another, and probabilities are no more nor less theory-laden than any other concepts we use in the description of nature. I agree with the first half of this answer and in general I have no quarrel with using theoretical language as a direct description of the empirical world. My suspicion is connected with a project that lies beyond the end of this book. Yet it colours the arguments throughout, so I think it should be explained at the beginning.

The kinds of probability that are being connected with causality in this chapter and in later chapters, and that have indeed been so connected in much of the literature on probabilistic causality over the last twenty years, are probabilities that figure in laws of regular association. They are the non-deterministic analogue of the equations of physics; and they might be compared (as I have done) to the regularities of Hume: Hume required uniform association, but nowadays we settle for something less. But they are in an important sense different from Hume's regularities. For probabilities are modal[30] or nomological and Hume's regularities were not. We nowadays take for granted the difference between a nomological and an accidental generalization: 'All electrons have mass' is to be distinguished from

---

[30] Cf. the discussion of this by B. van Fraassen in *The Scientific Image* (Oxford: Clarendon Press, 1983).

'All the coins in my pocket are silver'. The difference between probabilities and frequencies is supposed to be like that.

The reason I am uneasy about taking probabilities as an empirical starting-point is that I do not believe in these nomological regularities, whether they are supposed to hold for 100 per cent of cases or for some other percentage. I do not see many around, and most of those I do see are constructions of the laboratory. The more general picture I have in view takes the capacities which I argue for in this book not just to stand alongside laws, to be equally necessary to our image of science, but rather to eliminate the need for laws altogether. Capacities are at work in nature, and if harnessed properly they can be used to produce regular patterns of events. But the patterns are tied to the capacities and consequent upon them: they do not exist everywhere and every when; they are not immutable; and they do not license counterfactuals, though certain assumptions about the arrangement of capacities may. They have none of the usual marks they are supposed to have in order to be counted as nomologicals; and there is no reason to count them among the governors of nature. What makes things happen in nature is the operation of capacities.

This is a strong overstatement of an undefended view, and does not really belong here. But I mention it because it helps put into perspective the arguments of this book. For the doctrines I am opposing here are not so much those of Hume, but doctrines like those of Russell and Mach or, more in our own time, of Hempel and Nagel. Their philosophies admit laws of nature, so long as they are laws of association and succession; but they eschew causes. That seems to me to be a crazy stopping-point. If you have laws you can get causes, as I argue in this chapter. And you might as well take them. They are not much worse than the laws themselves (as I argue in Chapter 3), and with them you can get a far more powerful picture of why science works, both why its methods for finding things out make sense and why its forecasts should ever prove reliable.

Appendix: Back Paths and the Identification of Causes

Suppose there is a set of factors x_i, $i = 1,...,m$, each of which may, for all that is known, be a true cause of x_e, and such that

$$S: x_e = \Sigma a_i x_i,$$

where each x_i has an open back path. Assume transitivity of causality: if $(A \rightarrow B$ and $B \rightarrow C)$ then $(A \rightarrow C)$ ('$X \rightarrow Y$' means here, 'X causes Y'). Assume in addition a Reichenbach-type principle: given any linear equation in the form $z_e = \Sigma b_i z_i$, which represents a true functional dependence, either (*a*) every factor on the right-hand side is a true cause of the factor on the left-hand side, or (*b*) the equation can be derived from a set of equations for each of which (*a*) is true. Then each x_i is a true cause of x_e.

The argument is by *reductio ad absurdum*. Assume there is some true causal equation for x_e:

$$T: x_e = \Sigma b_j y_j \quad j = 1,...,n$$

and a set of true causal equations which allows $x_e - \Sigma b_j y_j = 0$ to be transformed into $x_e - \Sigma a_i x_i = 0$. Order the equations, including T, by the time index of the effects. The matrix of coefficients will then form a triangular array. Designate the equation in which ϕ is the effect by $\bar{\phi}$. Consider the x_i with the last time index which is not a true cause of x_e, and designate it 'x_1'.

Claim. It is necessary to use $\bar{x}_1$ to transform T to S.

Argument. It is only possible to introduce x_1 into T by using some equation in which x_1 occurs. This requires using either $\bar{x}_1$ itself or else some other equation, $\bar{\phi}$, such that ϕ occurs after x_1 and $x_1 \rightarrow \phi$. If $x_1 \rightarrow \phi$, then $\phi \neq y_j$ for any j, otherwise by transitivity x_1 is a true cause of x_e. Thus $\bar{\phi}$ introduces one new variable, ϕ, not already present in either T or S; and ϕ can only be eliminated by the use of a further equation for a variable with a later index. Only one of the $\bar{y}_j$ can eliminate ϕ without introducing yet another new variable needing to be eliminated. But ϕ can only be eliminated by $\bar{y}_j$ if ϕ appears in $\bar{y}_j$. This means $\phi \rightarrow y_j$—which is impossible by transitivity since $x_1 \rightarrow \phi$. Using an equation for a different variable ϕ' cannot help, for the same reason. The only way to eliminate ϕ' without introducing yet another variable itself in need of elimination is to use a $\bar{y}_j$. But if ϕ appears in $\bar{\phi}'$, ϕ' cannot appear in any $\bar{y}_j$, since then $(x_1 \rightarrow \phi)$ implies $(\phi' \rightarrow y_j)$. And so on. So the equation $\bar{x}_1$ itself must be used.

The equation $\bar{x}_1$ will introduce at least one factor—call it 'u_1'—that is part of an open back path. Hence $u_1 \neq x_i$ for all i, and also $u_1 \neq y_j$ for all j.

Claim. It is necessary to use $\bar{u}_1$ to transform T to S.

Argument.   Any equation for a variable after u_1 in which u_1 appears will introduce some new variable, ϕ, which must in turn be eliminated, since for all j, it is not possible that $u_1 \rightarrow y_j$, and also not possible for any i that $u_1 \rightarrow x_i$. This is because anything that u_1 causes must be *known* not to cause x_e. But ϕ itself cannot be eliminated without introducing yet another new variable, since ϕ cannot appear in any $\bar{x}_i$ or $\bar{y}_j$, otherwise by transitivity $u_1 \rightarrow x_i$ or $u_1 \rightarrow y_j$. And so on.

Now the conclusion follows immediately. For $\bar{u}_1$ introduces at least one variable, u_1', which in turn can only be eliminated by using $\bar{u}_1'$, and this equation introduces a u_1'' of which the same is true; and so on. Thus there will never be a finite set of equations by which S can be derived from T.

2

No Causes In, No Causes Out

2.1. Introduction

Chapter 1 showed how to get from probabilities to causes. It treated ideas that are primarily relevant to fields in the behavioural sciences and to quality control, and it tried to answer the question, 'How can we infer causes from data?' Section 2.2 of this chapter does the opposite. It starts with physics and asks instead the question, 'How can we infer causes from theory?' The principal thesis of section 2.2 is that we can do so only when we have a rich background of causal knowledge to begin with. There is no going from pure theory to causes, no matter how powerful the theory.

Section 2.3 will return to probabilities, to make the same point. It differs from Chapter 1, where the arguments borrowed some simple structures from causal modelling theory, especially the modelling theory of econometrics. Section 2.3 proceeds more intuitively, and its presentation is more reflective of the current discussion in philosophy of science. Its aim is to repeat the lesson of section 2.2, this time for the attempt to go from probabilities to causes. Its conclusion is one already implicit in Chapter 1: again, no causes in, no causes out.

This could be a source of despair, especially for the Humean who has no concept of causality to use as input. Section 2.4 reminds us that this is ridiculous. We regularly succeed in finding out new causes from old, and sometimes we do so with very little information in hand. Clever experimental design often substitutes for what we do not know. Section 2.5 returns to defend the thesis that some background causal knowledge is nevertheless necessary by showing why an alternative proposal to use the hypothetico-deductive method will not work. The chapter ends with a comparison of the informal arguments of section 2.2 with the more formally structured ones of Chapter 1.

2.2. Causes at Work in Mathematical Physics

Any field in which the aim is to tell adequate stories about what happens in nature will inevitably include causal processes in its descriptions. Though philosophical accounts may suggest the opposite, modern physics is no exception. Philosophers have traditionally been concerned with space and time, infinity, symmetry, necessity; in modern physics they have tended to concentrate on questions surrounding the significance of relativity theory, or parity conservation and the algebraic structure of quantum mechanics, rather than how physics treats specific physical systems. The case treated in this section will illustrate one kind of causal reasoning in physics, reasoning that attempts to establish causes not by doing experiments but by using theory. Specifically, the example shows how a theoretical model for laser behaviour helped solve a puzzle about what produces a dip in intensity where it was not expected.

Once one begins to focus on studies where causality matters, all the questions of the last chapter intrude into physics. Physics is mathematical; yet its causal stories can be told in words. How does the mathematics bear on physics' causal claims? This is precisely the question asked in Chapter 1; and the answer is essentially the same: functional dependencies of the right kind are necessary for causation, though they are not by themselves sufficient. Yet in certain felicitous cases they may combine with other facts already well established to provide sufficient conditions as well.

Renaissance thought assumed that mathematics was the provenance of real science, and that in real science the steps of a derivation mirror the unfolding of nature. This is not true of modern physics. In contemporary physics, derivations need not provide maps of causal processes. A derivation may start with the basic equations that govern a phenomenon. It may be both accurate and realistic. Yet it may not pass through the causes. This is exactly what happens in the laser example of this section. The derivation provided is highly accurate, but it does not reflect the right causal story.

Nevertheless, it appears that a derivation is necessary if the theory is to support the causal story, although it is not the derivation itself that provides the support. What is needed is a kind of back-tracking through the derivation, following to their origins the features that are mathematically responsible for the effect. Still, this special kind of back-tracking is not sufficient for causal support. This is not surprising if the derivation through which one is tracing did not go via the causes in the first place. In the example here, two structurally similar trackings are apparent, one which leads back to the causes, and one which does not.

Why are the features targeted by one tracing correct, while those targeted by the other are not? The answer in this case is that one fitted nicely into a causal process already fairly well understood, and the other could find no place. Here, quite clearly, background causal knowledge combined with new theoretical developments to fix the cause of the unexpected dip. The example is a particularly nice one because it has a familiar structure: it is a case of spurious correlation in a deterministic setting, just like Mackie's example of the Manchester hooters, only in this case the example is real and the regularities are given by the equations of a sophisticated quantum-mechanical model. Let us turn now to the example itself.[1]

The Lamb dip occurs in gas lasers. It is named after Willis Lamb, who first predicted its occurrence. Fig. 2.1, taken from Anthony Siegman's text *Lasers*, shows the elements of a typical laser oscillator. There are three basic parts: the laser medium; the pumping process; and mirrors or other devices to produce optical feedback. At thermal equilibrium most atoms are in the ground state rather than in any of the excited states. A laser works by pumping the medium until there is a population inversion—more atoms in an upper state than in a lower. Light near the transition frequency between the upper and the lower states will be amplified as the atoms in the upper state de-excite, and if the mirrors at the end of the cavity are aligned properly, the signal will bounce back and forth, reamplifying each time. Eventually, laser oscillation may be produced. Later I shall use this as a prime example of a capacity ascription in physics: an inverted population has the capacity to lase.

The Lamb dip occurs in the graph of laser intensity versus cavity frequency, as illustrated in Fig. 2.2. The atoms in the cavity have a natural transition frequency, ω; the cavity also has a natural frequency, v, depending on the placement of the mirrors. Prima facie it seems that the intensity should be greatest when the cavity frequency matches the atomic transition. Indeed, Lamb reports,

[1] A more detailed study of this case can be found in N. Cartwright, 'Causation in Physics: Causal Processes and Mathematical Derivations', in P. Asquith (ed.), *PSA [proceedings of the biannual Philosophy of Science Association meetings] 1984*, ii (East Lansing, Mich.: Philosophy of Science Association, 1985), 391–404.

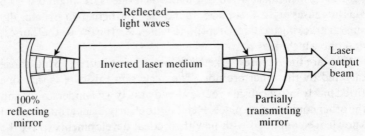

FIG. 2.1   *Laser oscillation*
Source: A. Siegman, *Lasers* (Mill Valley, Calif.: University Science Books, 1986), 4.

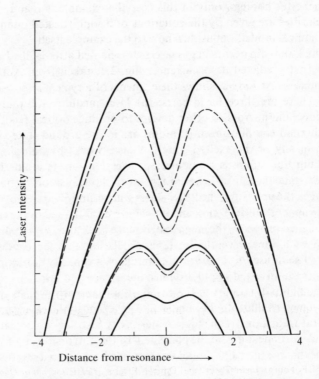

FIG. 2.2   *Predicted Lamb dip profiles at different pumping levels above threshold. Solid lines: simplified theory; broken lines: more exact analysis.*
Source: M. Sargent, M.O. Scully, and W.E. Lamb, *Laser Physics* (New York: Addison Wesley, 1977), s. 10.1.

I naively expected that the laser intensity would reach a maximum value when the cavity resonance was tuned to the atomic transition frequency. To my surprise, it seemed that there were conditions under which this would not be the case. There could be a local minimum, or dip, when the cavity was tuned to resonance [*i.e. cavity frequency = transition frequency*]. I spent a lot of time checking and rechecking the algebra, and finally had to believe the results.[2]

Lamb did not know it at the time, but the Lamp dip is caused by a combination of *saturation*, with its consequent *hole-burning*, and *Doppler shifting*, which occurs for the moving atoms in a gas laser such as helium-neon.

The concept of hole-burning comes from W.R. Bennett, and it was Bennett who first put hole-burning and the Lamb dip together in print; but both a footnote in the Bennett paper and remarks of Lamb (conversation, 1 October 1984) suggest that the connection was first seen by Gordon Gould. Bennett had been using hole-burning to explain unexpected beat frequencies he had been seeing in helium-neon lasers at the Bell Laboratories in 1961. But, Bennett explains, 'Ironically, a much more direct proof of the hole-burning process' is provided by the Lamb dip.[3]

Bennett's paper appeared in *Applied Optics* in 1962. Lamb's paper was circulating at the time but was not finally published until 1964. In fact, Lamb had been working on the calculations from the spring of 1961, and he says that he had already seen the dip (which Lamb calls 'the double peak' after the humps rather than the trough) by the fall of 1961.[4] Lamb wrote both to Bennett and to A. Javan about the prediction. Bennett, who had been measuring intensity versus cavity-tuning frequency in the helium-neon laser, sent back a tracing of only a single peak. Javan answered more favourably, for he had been seeing frequency-pushing effects that could be easily reconciled with Lamb's general treatment. Javan then did a direct experiment to show the dip, which he published later with A. Szoke.[5]

[2] W.E. Lamb, 'Laser Theory and Doppler Effects', *IEEE Journal of Quantum Electronics*, 20(6) (1984), 553.
[3] W.R. Bennett, 'Gaseous Optical Masers', *Applied Optics*, 1 (1962), supplement, 58.
[4] W.E. Lamb, Jr., conversation, 1 Oct. 1984.
[5] A. Javan and A. Szoke, 'Isotope Shift and Saturation Behaviour of the 1.15-MV Transition of NE', *Physical Review*, 10 (1963), 521 n. 12.

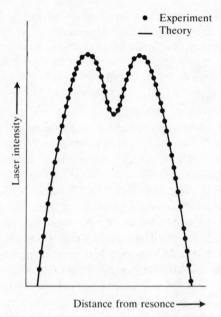

FIG. 2.3   *Carefully measured Lamb dip in a helium-neon laser*
Source: Siegman, op. cit., p. 1205.

The first published report of the dip was by R. A. McFarlane,[6] who attributed earlier failures to see the dip to the use of natural neon, whose two isotopes confound the effect. McFarlane used a single isotope instead, and got the results shown in Fig. 2.3.

Thus Lamb worked on the paper for three years before it was published. He is in general methodical and slow to publish. But there was special motivation in this case for holding back: he did not know what caused the dip. He could predict it, and he knew it existed; but he did not know what physical process produced it. This raises the first philosophical point of the example: the mathematical derivation of an effect may completely side-step the causal process which produces the effect; and this may be so even when the derivation is both (*a*) faultless and (*b*) realistic.

(*a*) Lamb's mathematical treatment was accurate and careful.

[6] R. A. McFarlane, W. R. Bennett, and W. E. Lamb, 'Single Mode Tuning Dip in the Power Output on an He-Ne Optical Maser', *Applied Physics Letters*, 2 (1963), 189–90.

Bennett described it as 'an extremely detailed and rigorous development of the theory of optical laser oscillation',[7] and that is still the
opinion today. In fact, Lamb's study of gas lasers was the first full
theoretical treatment of any kind of laser, despite the fact that Javan
had produced a gas laser at the end of December 1960, and that ruby
lasers had been operating since July 1960. The work of Schawlow
and Townes, which was so important for the development of lasers,
used bits of theory but gave no unified treatment.

(*b*) The calculations are based on a concrete, realistic model of the
gas laser. This contrasts, for example, with an almost simultaneous
theoretical treatment by Hermann Haken, which is highly formal
and abstract.[8] Lamb's calculations refer to the real entities of the
laser—moving gas molecules and the electromagnetic field that they
generate; and the equations govern their real physical characteristics, such as population differences in the atoms and the polarization
from the field. Nevertheless, the derivation fails to pass through the
causal process. Exactly how this happened will be described below.
Here the point to notice is that the failure to reveal the causes of the
Lamb dip did not arise because the derivation was unsound nor
because it was unrealistic.

Turn now to the concepts of hole-burning and saturation. For
simplicity, consider two level atoms with a transition frequency ω.
Once a population inversion occurs, a signal near the transition frequency will stimulate transitions in the atoms. The size of the
response is proportional both to the applied signal and to the population difference. The stimulated emission in turn increases the signal,
thereby stimulating an even stronger response. The response does
not increase indefinitely because the signal depopulates the upper
level, driving the population difference down, until a balance is
achieved between the effects of the pumping and the signal. This is
called *saturation* of the population difference. Oscillation begins
when the gain of the beam in the cavity is enough to balance the
losses due to things like leakage from the cavity. The intensity of the
oscillations builds up until the oscillation saturates the gain, and
brings it down. Steady-state oscillation occurs when the saturation
brings the gain to a point where it just offsets the losses.

---

[7] Bennett, op. cit., p. 58.
[8] H. Haken and H. Sauerman, 'Frequency Shifts of Laser Modes in Solid State and
Gaseous Systems', *Zeitschrift für Physik*, 176(1) (1963), 47.

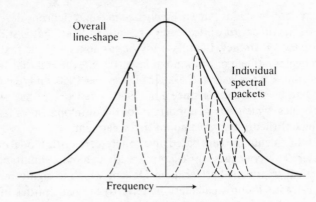

FIG. 2.4   *Composition of the Doppler-broadened line*
Source: Siegman, op. cit., p. 159.

Saturation produces unexpected effects when the laser medium is gas. In a gas laser the atoms are moving, and frequencies of the emitted light will be scattered around the natural transition frequency of the atoms, so that the observed spectral line will be much broader than the natural line for the atoms. That is because of Doppler-shifting. The moving atom sees a signal as having a different frequency from that of the stationary atom. The broadened line is actually made up of separate spectral packets with atoms of different velocities, where each packet itself has the natural line width, as in Fig. 2.4. This gives rise to the possibility of *hole-burning*: an applied signal will stimulate moving atoms whose effective transition frequency approximates its own frequency, but it will have almost no effect on other atoms. So the chart of the population difference versus frequency across the Doppler-broadened line shows a 'hole' in the population difference of the atoms near the applied frequency (Fig. 2.5).

The discussion so far has involved the Doppler shift due to the interaction of a moving atom with a single travelling wave. In a laser cavity there are two travelling waves, oppositely directed, which superpose to form a single standing wave. So the standing wave interacts with two groups of atoms—those whose velocities produce the appropriate Doppler-shifted frequency to interact with the forward wave and those whose Doppler-shifted frequency will interact with the backward one. These atoms have equal and opposite velocities. The holes pictured in Fig. 2.6 result.

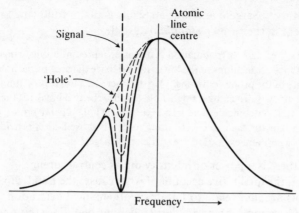

FIG. 2.5 *Burning a hole in a broadened line*
Source: Siegman, op. cit., p. 1173.

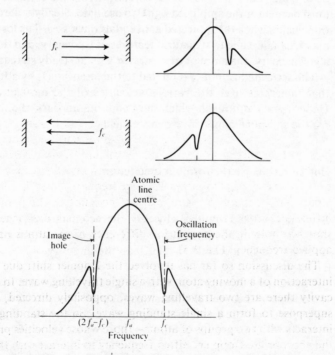

FIG. 2.6 *Travelling waves and resulting velocity holes in a standing-wave laser oscillator*
Source: Siegman, op. cit., p. 1201.

As Murray Sargent and Marian Scully explain, following Lamb's treatment in their *Laser Handbook* article,

> the holes, i.e., lack of population difference, represent atoms which have made induced transitions to the lower state. Hence the area of the hole gives a measure of the power [or intensity] in the laser. . . . For central tuning, the laser intensity is driven by a single hole because the two holes for the [velocities] v and $-v$ coincide. The area of this single hole [*i.e. the power*] can be less than that for the two contributing to detuned oscillation provided the Doppler width and excitation are sufficiently large.[9]

Hence there is a power or intensity dip at central tuning.

This is the qualitative account of what causes the Lamb dip. How does this account relate to Lamb's mathematical derivation? Here again the example is a particularly felicitous one, since this question is explicitly taken up in the advanced-level textbook by Sargent, Scully, and Lamb.[10] They begin by considering the conditions that must be met for the amplified signal to oscillate. Steady-state oscillation occurs when the saturated gain equals the losses. The losses, we may recall, are due to structural features of the cavity, and thus, for a given cavity, the amount of saturated gain at steady state is fixed.

The saturated gain (α_g) is related to the intensity (I) by a formula that integrates over the various velocities of the moving atoms. Under the conditions in which the Lamb dip appears, the formula for the saturated gain off resonance yields

$$\alpha_g^{0\,\text{ff}} \approx 1/(1 + 1/2 I^{0\,\text{ff}})$$

But on resonance the formula is different:

$$\alpha_g^{0\,\text{n}} \approx 1/(1 + I^{0\,\text{n}})$$

Since α_g is fixed by the physical characteristics of the cavity $\alpha_g^{\text{on}} = \alpha_g^{\text{off}}$, which implies that $I^{\text{on}} = 1/2 I^{\text{off}}$, that is, the intensity on resonance is significantly smaller than the intensity off resonance. This is the Lamb dip.

The causes of the dip can be discovered by finding the source of the difference between the formulae for α_g^{on} and α_g^{off}. Most immediately the difference comes from the denominator of the full

---

[9] M. Sargent and M. Scully, 'Theory of Laser Operation: An Outline', in F. T. Arecchi and E. O. Schulz (eds.), *Laser Handbook* (Amsterdam: North-Holland, 1972), ii. 80.

[10] Sargent *et al.*, *Laser Physics*, ch. 10.1.

formula for α_g before integration. The denominator is a function of two Lorentzian functions (the Lorentzian is defined by $L_\gamma(x) = (\gamma^2/\gamma^2 + \omega^2)$). One Lorentzian has $+ \nu$ in its ω argument; the other, $- \nu$. When the integration over ν is carried out only one of the Lorentzians contributes off resonance, but both contribute on resonance.

This is just what Sargent, Scully, and Lamb note:

The Lorentzians . . . show that holes are burned in the plot of [the intensity curve]. Off resonance ($\nu \neq \omega$), one of the Lorentzians is peaked at the detuning value $\omega - \nu = K\nu$, and one at $\omega - \nu = - K\nu$, thereby burning *two* holes . . . On resonance ($\nu = \omega$), the peaks coincide and a *single* hole is burned.[11]

The next step is to discover how the two different Lorentzians got there in the first place. The answer is found just where it is to be expected. Laser intensity is due to depopulating the excited level. Populations are depleted at two different velocities because there are two running waves with which the atoms interact. So the source of the Lorentzians should be in the running waves—as indeed it is. Sargent, Scully, and Lamb know this from comparing equations. The electromagnetic field in the cavity represented by a standing wave is really composed of two running waves. The Lorentzians first appear when the formula for the standing wave is written as a sum of the two running waves. Sargent, Scully, and Lamb report:

For $\nu > \omega$, an atom moving along the z axis sees the first of the running waves . . . 'stretched' out or Doppler downshifted. . . . Comparison of the equation [*which writes the standing wave as a sum of the two running waves*] with [*the equation used to generate the population difference*] reveals that the Lorentzian in [*the saturation factor*] results from this running wave. . . . Similarly, an atom moving with velocity $- \nu$ sees the second standing wave down-shifted, interacts strongly if the atom with velocity ν did, and produces the [*second*] Lorentzian.[12]

Thus the story is complete. The Lamb dip has found its source in the combination of saturation and Doppler broadening, as promised.

In this exposition Sargent, Scully, and Lamb trace the mathematics and the causal story they want to tell in exact parallel. Fig. 2.7 summarizes the two, side by side. Notice exactly what Sargent, Scully, and Lamb do. They do not lay out a derivation; that has been

---

[11] Ibid. 149.
[12] Ibid. 150.

Off resonance: $a \rightarrow b = b$ is derived from a	*Off resonance*: $a \Rightarrow b = a$ causes b
Each of the 2 running-wave terms $\rightarrow 1$, each, of the 2 Lorentzians in the saturation factor	Existence of waves running in 2 directions $\Rightarrow$ saturation of the 2 velocities for which the 2 oscillating frequencies are appropriately up-and down-shifted
$\downarrow$	$\Downarrow$
Each of the Lorentzians in the saturation factor $\rightarrow$ a Lorentzian (in the denominator) that reduces the population difference at the appropriate Doppler-shifted velocity	Saturation in each of 2 velocities $\Rightarrow$ reduction in the population difference at those 2 velocities
$\downarrow$	$\Downarrow$
2 Lorentzians in the denominator of the population difference $\rightarrow$ 2 Lorentzians in the denominator of the gain formula $\rightarrow$ greater intensity in the cavity	Reduction in the population difference at 2 velocities (and hence a greater number of atoms which have made induced transitions to the lower state) $\Rightarrow$ greater intensity in the cavity

FIG. 2.7 *Comparison of the causal account and the mathematical derivation*
Source: N. Cartwright, 'Causation in Physics: Causal Processes and Mathematical Derivations', in P. Asquith (ed.), *PSA 1984*, ii (East Lansing, Mich.: Philosophy of Science Association, 1985), 400.

done earlier. Rather, they take an intelligent backwards look through the derivation to pick out the cause. First, they isolate the mathematical feature that is responsible for the exact characteristic of the effect in question—in this case, the fact that two Lorentzians contribute off resonance and only one on resonance; second, they trace the genealogy of this feature back through the derivation to its origin—to the two terms for the oppositely directed running waves; third, they note, as they go, the precise mathematical consequences that the source terms force at each stage. The mathematics *supports the causal story* when these mathematical consequences traced back through the derivation match stage by stage the hypothesized steps of the causal story. This is the kind of support that causal stories need, and until it has been accomplished, their theoretical grounding is inadequate.

Although this is just one case, it is not an untypical case in physics. Just staying within this particular example, for instance, one could easily lay out a similar retrospective mathematical tracing for the causal claim that an applied beam saturates the population difference. For contrast, one might look at Bennett's own 'hole-burning model'[13] which does not (so far as I can reconstruct) allow the kind of backwards causal matching that Lamb's does. I am not going to do that here, but simply summarize my conclusion: no matter how useful or insightful the more qualitative and piecemeal considerations of Bennett are, they do not provide rigorous theoretical support for the causal story connecting saturation and Doppler-broadening with the Lamb dip. Only matching back-tracking through the derivation in a realistic model can provide true theoretical support for a causal hypothesis.

Although this kind of back-tracking is necessary, it is not sufficient. For a derivation may well predict a phenomenon without mentioning its causes. This is just what happened in Lamb's original work. He predicted the dip, but missed its causes. When he went back to look for them, he found something in his derivation that was mathematically necessary for the dip, but which did not cause it. Just like the dip, the feature he focused on was the result of the Doppler-shifting that occurs in a gas laser, but it itself has no direct effect on the dip. For a while Lamb was misled by a spurious correlation.

To see this point, one needs to know something about the role of electric dipole moments, or atomic polarizations, in Lamb's theory. Lamb uses what he calls a 'self-consistency' treatment. He begins by modelling the radiating atoms of the laser medium as dipole oscillators; the amount of radiation is dependent on the dipole moment. The atoms start to radiate in response to the field in which they are located, and they in turn contribute to the field. The contribution can be calculated by summing the individual dipole moments, or

---

[13] Bennett, op. cit.; id., 'Interactions on Gas–Laser Transitions', *Physical Review*, 18 (1967), 688.

atomic polarizations, to get a macroscopic polarization which then plays its normal role in Maxwell's equations. If the laser reaches steady-state oscillation, this process comes to a close. The field which results from Maxwell's equations must then be the same as the field which causes the atoms to radiate. As Lamb explains:

We understand the mechanism of laser oscillation as follows: an assumed electromagnetic field $E(r,t)$ polarizes the atoms of the medium creating electric dipole moments $P_i(r,t)$ which add up to produce a macroscopic polarization density $P(r,t)$. This polarization acts further as the source of a reaction field $E'(r,t)$ according to Maxwell's equations. The condition for laser oscillation is then that the assumed field be just equal to the reaction field.[14]

Fig. 2.8 is Lamb's own diagram of the process.

Since the characteristics of laser oscillation that Lamb wanted to learn about depend on the macroscopic polarization, he was led to look at the atomic polarizations. In quantum mechanics the derivative for the atomic polarizations depends on the population differences, and vice versa. The two are yoked together: the rate of change of the population difference at a given time depends on the polarization at that time. But the polarization has been changing in a way dependent on the population difference at earlier times, which depends on the polarization at those times, and so on.

To solve the equations, Lamb used a perturbation analysis. Roughly, the perturbation analysis goes like this. To get the first-order approximation for the polarization, insert the *initial* value for the population difference, and integrate over time. Similarly, for the first-order approximation in the population difference, insert the initial value for the polarization. For the second-order approximations use, not the initial values in the time integrals, but the first-order values; and so on. It turns out that the contributions alternate through the orders, in just the way one naturally thinks about the problem. First, in zero order, the original inverted population difference interacts with the field, which produces a polarization contribution in first order; this in turn makes for a new population contribution in second order; which gives rise to a new polarization in third order. The first-order polarization, which has not yet taken into account any feedback from the stimulated atoms, is accurate enough to calculate the threshold of oscillation but not to study

---

[14] W. E. Lamb, 'Theory of an Optical Maser', *Physical Review*, 134 (15 June 1964), 1429.

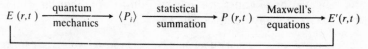

FIG. 2.8 *Summary of Lamb's derivation*
Source: W. E. Lamb, 'Theory of Optical Maser Oscillators', in C. H. Townes and P. A. Miles (eds.), *Quantum Electronics and Coherent Light* (New York: Academic Press, 1964).

steady-state oscillation. Hence Lamb concentrated on the third-order polarization.

Calculating the intensity from the third-order polarization, one discovers the Lamb dip. But Lamb did not see the cause for the dip. Why? Because of an unfortunate shift in the order of integration.[15] In calculating the third-order polarization one must integrate through time over the second-order population difference. Recall from the discussion of Doppler-shifting that the population difference varies significantly with the velocity of the atoms. Since the macroscopic polarization depends on the total contribution from all the atoms at all velocities, the calculation must integrate over velocity as well. For mathematical convenience, Lamb did the velocity integral first, then the time integral, and he thus wiped out the velocity information before solving for the population difference. He never saw the two holes at $+ v$ and $- v$ that would account for the dip.

By 1963 Gould and Bennett had suggested the hole-burning explanation, and Lamb had inverted the integrals and derived the velocity dependence of the population difference. The calculation of the intensity, and back-tracking of the kind done by Sargent, Scully, and Lamb, is routine from that point on. But in 1961 and 1962 Lamb had not seen the true causal story, and he was very puzzled. What did he do? He did exactly the kind of mathematical back-tracking that we have seen from the text he wrote later with Sargent and Scully. He himself says, 'I tried very hard to find [the] origin [of the dip] *in my equations*.'[16] Lamb's back-tracking is just what one would expect: 'The dependence of a typical term . . . is . . .'; 'the physical consequence of the appearance of terms involving Kv . . . is . . .'; 'only

---

[15] See Lamb, 'Laser Theory and Doppler Effects'; id., 'Theory of an Optical Maser', pp. A1448, A1449.

[16] Id., 'Laser Theory and Doppler Effects', p. 553 (my italics).

the first two possibilities are able to lead to non vanishing interference . . .' Finally Lamb concludes,

Physically, one may say that a dominant type of process involves three interactions: first, one with a right (left) running wave at t''', then one with a left
(right) running wave at t'', and finally one with a left (right) running wave at
t', with the time integrals obeying $t - t' = t'' - t'''$ so that the accumu-
lated Doppler phase angle . . . cancels out at time t'''.[17]

Lamb spent a good deal of time trying to figure out the physical
significance of these time interval terms but he could not find a
causal role for them:[18] 'I never was able to get much insight from this
kind of thing. The correct interpretation would have been obvious if
I had held back the v integration . . .'[19]

So here is a clear case. The mathematical back-tracking that I
claim is necessary to support a causal story is not sufficient to pick
out the causal story. The velocity dependence of the population
difference plays a significant role in the physical production of the
dip, whereas facts about the time intervals are merely side-effects.
Yet the mathematical dependencies are completely analogous. The
first is singled out rather than the second, not by mathematical back-
tracking, but by our antecedent causal knowledge, which in this case
is highly detailed. Lamb starts with a sophisticated causal picture,
outlined in Fig. 2.8—a picture of an applied field which polarizes the
atoms and produces dipole moments. The dipole moments add up to
a macroscopic polarization that produces a field which polarizes the
atoms, and so on. The velocity dependence fits in a clear and precise
way into this picture. But no role can be found for the time difference equalities. These find no place in the causal process that we
already know to be taking place.

The lesson of this example for physics is that new causal knowledge can be built only from old causal knowledge. There is no way
to get it from equations and associations by themselves. But there is
nothing at all special about physics in this respect. I have gone into
an example in physics in some detail just because this is one area in
which the belief has been particularly entrenched that we can make
do with equations alone. That is nowhere true. The last chapter
argued that facts about associations can help us to find out about

[17] Id., 'Theory of an Optical Maser', p. A1448.
[18] Id., conversation, 1 Oct. 1984.
[19] Id., 'Laser Theory and Doppler Effects', p. 553.

causes; but it must be apparent that the methods proposed there need causal information at the start. The next section will reinforce this point, this time setting the discussion of probabilities and causes in a more informal context than that of Chapter 1.

2.3. New Knowledge Requires Old Knowledge

It is an old and very natural idea that correlation is some kind of sign of causation, and most contemporary philosophical accounts take this idea as their starting-point: although causes may not be universally conjoined with their effects, at least they should increase their frequency. Formally, $P(E/C) > P(E/ \neg C)$. But as Chapter 1 made clear, a correlation between two variables is not a reliable indicator that one causes the other, nor conversely. Two features A and B may be correlated, not because either one causes the other, but because they are joint effects of some shared cause C. We have just seen an example of this in the last section, where the Lamb dip and the time-interval equalities owed their co-association to the combination of hole-burning and Doppler-shifting. The correlation described in Chapter 1 between Mackie's Manchester hooters and the stopping of work for the day in London is another good illustration. In these kinds of case, the frequency of Bs among As is greater than among not-As for a somewhat complicated reason. A case in which A occurs is more likely to have been preceded by C than one in which A does not occur, and once a C has been selected, the effect B will be more likely as well.

The conventional solution to this problem is to hold C fixed: that is, to look in populations where C occurs in every case, or else in which it occurs in no case at all. In either population, selecting an A will have no effect on the chance of selecting a C; so if A and B continue to be correlated in either of those populations, there must be some further reason. If *all* the other causes and preventatives of B have been held fixed as well, the only remaining account of the correlation is that A itself is a cause of B. So to test for a causal connection between a putative cause C and an effect E, it is not enough to compare $P(E/C)$ with $P(E/ \neg C)$. Rather one must compare $P(E/C \pm F_1 \ldots \pm F_n)$ with $P(E/ \neg C \pm F_1 \pm \ldots \pm F_n)$, for each of the possible arrangements of E's other causes, here designated by $F_1, \ldots, F_n$. The symbol $\pm F_n$ indicates a definite choice of either F_n or $\neg F_n$. This suggests the following criterion:

CC: *C* causes *E* iff
$$P(E/C \pm F_1 \pm \ldots \pm Fn) > P(E/\neg C \pm F_1 \ldots \pm F_n),$$
where $\{F_1, \ldots, F_n, C\}$ is a complete causal set for *E*.

To be a complete causal set for *E* means, roughly, to include all of *E*'s causes. The exact formulation is tricky; at this stage one should attend to more evident problems.

The practical difficulties with this criterion are conspicuous. The conditioning factors $F_1, \ldots, F_n$ must include every single factor, other than *C* itself, that either causes or prevents *E*, otherwise the criterion is useless. Finding a correlation, or a lack of it, between *C* and *E* signifies nothing about their causal connection. A sequence of studies in econometrics will illustrate. The studies, published between 1972 and 1982 by Christopher Sims, used a criterion due to C. W. J. Granger to test the Keynesian hypothesis that money causes income.[20] The Granger criterion is the exact analogue of condition *CC*, extended to treat time-series data.

Sims began his discussion of causal ordering with the remark, 'It has long been known that money stock and current dollar measures of economic activity are positively correlated.' But which way, if any, does the causal influence run? Sims used a test based on the Granger definition to decide: 'The main empirical finding is that the hypothesis that causality is unidirectional from money to income agrees with the postwar U.S. data, whereas the hypothesis that causality is unidirectional from income to money is rejected.'[21]

In this study no variables other than money stock and gross national product were considered. But in a four-variable model, containing not just money stock and GNP but also domestic prices and nominal interest rates, the conclusion no longer followed. Without interest rates, changes in money supply account for 37 per cent of the variance in production; with interest included, the money supply accounts for only 4 per cent of variance in production.[22] When the model is expanded to include federal expenditures and federal

---

[20] According to Granger, the past and current values of a variable *Y* cause the current value of the variable *X*, *relative to past and current value of Z*, just in case including information about the *Y* values changes the probability for *X*, holding fixed both the history of *X* itself and all past and current values of *Z*.

[21] C. A. Sims, 'Money, Income, and Causality', *American Economic Review*, 62 (1972), 540.

[22] Id., 'Comparison of Interwar and Postwar Business Cycles: Monetarism Reconsidered', *American Economic Review*, 70 (1980), 250–7.

revenues as well, the results are slightly different again. Production then seems to be influenced by both revenues and interest rates; and money supply by interest rates alone, with no direct influence from money to production. According to Sims, the 'simultaneous downward movements in [money] and [production] emerge as responses to interest rate disturbances in the larger system'.[23] So the hypothesis that money causes income is not supported in this model. The well-known correlation between money stock and production appears to result from the action of a joint preventative. But the question remains, is a six-variable model large enough? Or will more variables reverse the findings, and show that money does cause activity in the economy after all?

The question of which variables to include is not only practical but of considerable philosophical importance as well. Followers of Hume would like to reduce causal claims to facts about association. Granger's definition is an attempt to do just that. But, like the related criterion CC, it will not serve Hume's purposes. The definition is incomplete until the conditioning factors are specified. How are these factors to be selected? Roughly, every conditioning factor must be a genuine cause (or preventative) and together they must make up a complete set of causes. Nothing more, nor less, will do. So neither CC nor the definition of Granger can provide a way to replace the concept of causality by pure association, a thesis already suggested by Chapter 1.

Granger's own solution to the problem is to include among the conditioning factors 'all the knowledge in the universe available at that time' except information about the putative cause.[24] But this suggestion does not work. For it takes a strategy that is efficacious in the analysis of singular causal claims and tries to apply it to claims at the generic level, where it does not make sense. The question at issue is not one of singular causation: 'Did the occurrence of C in a specific individual at a specific time (say 1000 hrs. on 3 March 1988) cause an occurrence of E after a designated interval (say at 1100 hrs. on 3 March 1988)?'; but rather the general one: 'Is it a true law that Cs occurring at some time t cause Es to occur at $t + \delta t$?'

---

[23] Id., 'Policy Analysis with Econometric Models', in W.C. Brainaud and G.L. Perry (eds.), *Brookings Papers on Economic Activity*, 4 (Washington, DC: Brookings Institution, 1982), 135.

[24] C.W.J. Granger, 'Testing for Causality: A Personal Viewpoint', *Journal of Economics, Dynamics and Control*, 2 (1980), 335.

The first claim refers to a specific individual at a specific time, and that individual will have some fixed history. One can at least imagine the question, 'For individuals with just that history, what is the probability for E to occur given that C does?' But in the second case, no specific individual and no specific time are picked out. No history is identified, so the suggestion to hold fixed everything that has happened to that individual up to that time does not make sense. What is needed is a list that tells what features are relevant at each time—not at each historical time but at each 'unit interval' before the occurrence of E. Indeed, this is probably the natural way to read Granger's suggestion: hold fixed the value of all variables from a designated set which have a time index earlier than that of the effect in question. But then the question cannot be avoided of what makes a factor relevant. If the correct sense of relevance is causal relevance, the problem of circularity remains.

A weaker, non-causal sense of relevance could evade the problem; but it is difficult to find a satisfactory one. The most immediate suggestion is to include everything which is statistically relevant, that is, everything which itself makes a difference to the probability of the effect. But this suggestion does not pick a unique set: F and F' can both be irrelevant to E, *simpliciter*: $P(E/F) = P(E/\neg F)$ and $P(E/F') = P(E/\neg F')$; yet each is relevant relative to the other: $P(E/F \pm F') \neq P(E/\neg F \pm F')$ and $P(E/F' \pm F) \neq P(E/\neg F' \pm F)$. Are F and F' both to be counted causes, or neither?

Sometimes it is maintained that probability arrangements like these will not really occur unless both F and F' are indeed genuine causes. They may appear to be based on the evidence of finite data; but a frequency in a finite population is not the same as a true probability on almost anyone's view of probability. True probabilities do not behave that way.[25] This seems an excellent proposal. But it does require some robust sense of 'true' probability; and one that does not itself depend on a prior causal notion, if it is to avoid the same circularity that besets principles like CC.

One ready candidate is the concept of personal, or subjective, probability. But this concept will not serve as it is usually developed. Much work has been done to fill out the unfledged idea of personal

[25] This is the view, I believe, of Brian Skyrms, and of Ellery Eells and Elliott Sober. Cf. B. Skyrms, *Causal Necessity*, (New Haven, Conn.: Yale University Press, 1980); E. Eells and E. Sober, 'Probabilistic Causality and the Question of Transitivity', *Philosophy of Science*, 50 (1983), 35–57.

degree of belief, to develop it into a rich and precise concept. Notably there is the proof that degrees of belief that are coherent (in the sense that, translated into betting rates, they do not allow a sure-win strategy on the part of the betting opponent) will necessarily satisfy the probability calculus, as well as the identification of degrees of belief with betting rates; and the associated psychological experiments to determine, for instance, how fine-grained people's degrees of belief actually are. But nothing in this work tailors degrees of belief to match in the right way with the individual's causal beliefs. The concept of degree of belief must be narrowed before it will serve to rule out renegade probabilities that make trouble for the analysis of causation. Again, the question is, can this be done by using only concepts which are philosophically prior to causal notions? My own view, as an empiricist, is that probabilities are just frequencies that pass various favoured tests for stability. In that case it is quite evident that problematic cases occur regularly, and some stronger sense of relevance than statistical relevance will be needed to infer causes from probabilities.

An example of precisely the sort just described has shaken Clark Glymour's confidence in his own bootstrap method, which I advocate throughout this book. In *Discovering Causal Structure*, Glymour and co-authors Richard Scheines, Peter Spirtes, and Kevin Kelly describe an experiment in which newly released felons were given unemployment payments for up to six months after getting out of prison so long as they were not working. The rearrest rate was about the same in the test group, which received the unemployment payment, as in the control group, which did not. Did that mean that the payments were causally irrelevant to rearrest? The experimenters thought not. They hypothesized that the payments did indeed reduce recidivism, but that the effect was exactly offset by the influence of unemployment, which acts to increase recidivism. Glymour and co-authors favour the opposite hypothesis, that the payments are irrelevant; correlatively, they must endorse the claim that so too is unemployment.

It is clear from the tone of the discussion that the joint authors take this stand in part because of the bad arguments on the other side. Indeed, they introduce the example in order to undercut a lot of well-justified objections to the way in which causal modelling is put to use: 'In criticisms of causal modelling, judgements about issues of *principle* are often entangled with criticisms of particularly bad practices.'[26] But the example also fits their central philosophical theses. They maintain that the relation between causal claims and statistics is hypothetico-deductive, so that there is never any way to infer from the statistics to the hypotheses. Nevertheless, among hypotheses that are all consistent with the same statistical data, some will be more probable than others.

The hypotheses that are more likely to be true for Glymour, Scheines, Kelly, and Spirtes are those that balance among three principles which 'interact and conflict'. One principle is called Thurstone's Principle: 'Other things being equal, a model should not imply constraints that are not supported by the sample data.'[27] This principle matters for the realistic and difficult job that is undertaken in *Discovering Causal Structure*—that of discovering the structures from actual data. In the language of Chapter 1 of this book, they simultaneously confront both Hume's problem—how do causes relate to probabilities?—and the problem of estimation—how can probabilities be inferred from very limited data? In the context of Hume's problem alone, Thurstone's Principle offers the relatively trivial advice to reject models that have false consequences; hence it will not be of special interest here.

The second of the three principles is very familiar—the Principle of Simplicity: simpler models are more likely to be true than complex ones, where 'In the special case of causal models, we understand simpler models to be those that posit fewer causal connections. In effect, we suppose an initial bias against causal connections, and require that a case be made for any causal claims.'[28] This explains their preference for the model which has taken neither cash-in-pocket nor unemployment to influence recidivism, rather than both.

The remaining criterion is expressed in Spearman's Principle: 'Other things being equal, prefer those models that, for all values of their free parameters (the linear coefficients), entail the constraints judged to hold in the population.'[29] To see how this principle works in the recidivism example, return to the methods of Chapter 1. Let r represent recidivism, c, cash-in-pocket, and u, unemployment. It is clear from the situation that unemployment is one of the causes that

[26] C. Glymour, R. Scheines, P. Spirtes, and K. Kelly, *Discovering Causal Structure* (New York: Academic Press, 1987), 32.

[27] Ibid. 100.

[28] Ibid. 101.

[29] Ibid. 100.

produces the cash grant: so set $c = u + w$. Adding the second hypothesis that neither c nor u acts as a cause of r gives rise to Model 1:

Model 1

$$c = \alpha u + w$$
$$r = v$$

Here v and w are error terms, assumed to be independent of the explicit causes in their respective equations. Coupling instead the hypothesis that both c and u are causes gives Model 2:

Model 2

$$c = \alpha u + w$$
$$r = \beta u + \gamma c + v$$

To produce the result in Model *2* that cash-in-pocket and unemployment exactly cancel each other in their influence on recidivism, set $\alpha\gamma = -\beta$. In this case both models imply that there will be no correlation between cash-in-pocket and recidivism. But they differ in the degree to which they satisfy Spearman's Principle. Model 1 has no free parameters (excepting those for the distributions of the error terms); the structure itself implies the observed lack of correlation. Model 2 implies this result only when an additional constraint on the free parameters is added. So according to Spearman's Principle, Model 1 is more likely to be true than Model 2.

The kind of practical empiricism that I advocate looks at the matter entirely differently. If unemployment really does cause recidivism, then, given the lack of correlation, cash-in-pocket must inhibit it; and if unemployment does not cause recidivism, then cash-in-pocket is irrelevant as well. The statistics cannot be put to work without knowing what the facts are about the influence of unemployment; and there is no way to know short of looking. Glymour, Scheines, Kelly, and Spirtes advocate a short cut. For them, it is more likely that unemployment does not cause recidivism than that it does. That is in part because of their 'initial bias against causal connections'. But the hypothesis that unemployment does not cause recidivism is as much an empirical hypothesis as the contrary; and it should not be accepted, one way or the other, unless it has been reliably tested. Failing such evidence, how should one answer the question, 'Does cash-in-hand inhibit recidivism?' Glymour, Scheines, Kelly, and Spirtes are willing to claim, 'Probably not'. But the empiricist who insists on measurement will say, 'We don't know; and no matter how pressing the question, it is no good pretending one has (or probably has) an answer, when one doesn't.' For an empiricist, there is no alternative to evidence.[30]

2.4. How Causal Reasoning Succeeds

The first main thesis of this book is that causal hypotheses pose no special problems for science. They can be as reliably tested as anything else. Chapter 1 showed one method that works; it explained how we can use probabilities as instruments to measure causes. But in the face of the discussion of sections 2.2 and 2.3, that claim seems disingenuous. Certainly we can measure causes with probabilities, but only if we have an impossible amount of information to begin with. It seems that a method that requires that you know all the other causes of a given effect before you can establish any one them is no method at all.

But that is not true. For the method does not literally require one to know all the other causes. Rather, what you must know are some facts about what the probabilities are in populations that are homogeneous with respect to all these other causes, and that you can sometimes find out without first having to know what all those causes are. That is the point of the randomized experiment which will be the first topic of this section. I will not spend long in discussing randomized experiments, however, for the theory is well known; and I want quickly to go on to an interesting example in physics where the experimenters claim they do know about all the other possible factors that might make a difference. That is the topic of the second section.

2.4.1. *The Randomized Experiment*

The classical discussion of randomization is R. A. Fisher's *Design of*

---

[30] It should be noted that this does not mean that no causal hypotheses can be ruled out until an experiment has been done. Empiricists too can have an 'initial bias' against all sorts of 'outlandish' causal hypotheses—so long as the reasons that make these hypotheses outlandish can be marshalled into a sound empirical argument that implies that the hypotheses are false. Often arguments of this type will turn out to employ empirical premises that are very vague but nevertheless exceedingly well established.

Experiments,[31] and his own example will illustrate. An experiment is conducted to test whether a particular person—let us call her *A*—can tell whether it is the milk or the tea which has first been added to the cup. *A* is to be given eight cups of tea to judge, four in which the tea came first, and four in which the milk came first. The null hypothesis in this case is that *A* possesses no special talents at all, and her sorting is pure chance—that is, it has no cause. But obviously there is a great deal of room between having no special talent and sorting purely by chance. There is a near-infinity of other differences between cups of tea that may cause *A* to sort them one way rather than another. What if all the cups with milk in first also had sugar in them? This is obviously an easy occurrence to prevent. But what about the others? Fisher says:

In practice it is probable that the cups will differ perceptibly in the thickness or smoothness of their material, that the quantities of milk added to the different cups will not be exactly equal, that the strength of the infusion of tea may change between pouring the first and the last cup, and that the temperature also at which the tea is tasted will change during the course of the experiment. These are only examples of the differences probably present; it would be impossible to present an exhaustive list of such possible differences appropriate to any one kind of experiment, because the uncontrolled causes which may influence the result are always strictly innumerable.[32]

Fisher points here to just the problem discussed in section 2.3. The putative cause of *A*'s choices is the fact that the milk and not the tea was first added to the cup. But so long as this fact is correlated with any other possible influences on her choice, the probabilities in the experiment will signify nothing about the hypothesis in question. Formula *CC* attacks this problem directly. The correlation between the putative cause and all other possible causes is broken by controlling for all these others. Fisher opposes this strategy, and for the obvious reason:

whatever degree of care and experimental skill is expended in equalizing the conditions, other than the one under test, which are liable to affect the result, this equalization must always be to a greater or less extent incomplete, and in many important practical cases will certainly be grossly defective.[33]

[31] (London: Oliver and Boyd, 1953).
[32] Ibid. 55.
[33] Ibid. 19.

The solution instead is to randomize.

An ideal randomized treatment-and-control group experiment must satisfy two related conditions. It will consist of two groups, the treatment group and the control group; in Fisher's experiment, the cups with milk in first and those with tea in first. The first requirement is that all other causes that bear on the effect in question should have the same probability distribution in both groups. Thick cups should not be more probable in one group than the other, nor strong infusions, nor any other possibly relevant factor. The second requirement is that the assignment of individuals to either the treatment or the control group should be statistically independent of all other causally relevant features that an individual has or will come to have. The random selection of individuals for one group or the other is supposed to be a help to both ends. But it is not the full story. Consider the placebo effect. How is the medicine to be introduced without also introducing some expectation for recovery, or perhaps, in the case of counter-suggestibility, some dread? This itself may be relevant to whether one gets better or not. There are a number of clever and well-known devices, for example blinds and double blinds, to deal with problems like these, and a vast amount of accompanying literature, both philosophical and practical, discussing their effectiveness.

I do not want to pursue this literature, but instead to return to my brief characterization of the ideal experiment. All of these procedures and devices are designed to ensure that the results of the real experiment will be as close as possible to the results of an ideal experiment. But one must still ask the prior question: what do the results of an ideal experiment signify? If the effect should have a higher probability in the treatment group than in the control group, what does that say about the causal powers of the treatment? In Chapter 3 I will describe an argument that starts with premises about the nature of causality—like 'Every occurrence of the effect has a cause'—and ends with the conclusion that the kind of probability increases prescribed by Principle CC will be sufficient to guarantee the truth of the corresponding causal law, and conversely. One could try a similar tactic with the ideal experiment: try to lay out just what assumptions must be made in order to justify the ideal experiment as an appropriate device for establishing causal laws. But this extra argument is not necessary. For it is not difficult to see that

that experiment—that is, ideal experiment—and Principle CC are bound to agree.

To show this properly one needs to provide some formal framework where the ideal experiment can be defined precisely and a proof can be given. I think the gain is not sufficient to warrant doing that here. Intuitively the result is easy to see by considering a mental partitioning of both the control group and the test group into the kinds of homogeneous population that appear in CC. If, as CC requires, the probability of the effect occurring is greater with the treatment than without it in each of these sub-populations, then it will also be greater in the whole population where the treatment occurs than where it does not; and the converse as well. The point is that CC and the ideal experiment dovetail. We do not have the puzzle of two separate and independent methodologies that are both supposed to establish the same facts. Instead we can see why one works if and only if the other does.

In fact, matters are more complicated, for a number of caveats have to be added. These are taken up in Chapter 3. The convergence of the two methodologies is only assured in certain especially simple cases, where a given cause has no contrary capacities—that is, capacities both to produce and to inhibit the effect. When more complicated cases are taken into account, the randomized experiment, even in the ideal, is not capable of picking out the causal laws correctly, whereas an amended version of CC is. That means that the methodology of CC is more basic than that of the controlled experiment. One can derive from CC why experiments allow us to draw some kinds of conclusion and not others; and also why they fail where they do.

In part it is because Principle CC is more fundamental that I concentrate so much on it in this book and on the more formal but analogous methods of causal modelling theories, rather than pursuing the study of randomized experiments. The point of discussing them here is to recall that the demand for total information that seems to follow from CC is not necessarily fatal. Sometimes we can find out what would happen were all the other causes held fixed without even knowing what the factors are that should be held fixed. It is important to keep in mind, however, that it takes an ideal experiment to do this, and not a real one. For, as with Principle CC itself, the connection between causality and regularity is drawn already well above the level of real data and actual experiment. It is not frequencies that yield causes, but probabilities; and it is not results in real experiments, where subjects are assigned to groups by a table of random numbers, but rather in ideal experiments where randomization is actually achieved.

2.4.2. The Totally Controlled Experiment

The technique of randomization just discussed is designed to deal with cases where we lack information. We have no clear idea what the background causes are, so we try to get by without that knowledge. Now I want to turn to an example which is diametrically opposite, where the plan is to determine exactly what the disturbing factors will be and to control every one of them. The example is the Gravity Probe-B experiment (GP-B) or the Stanford Relativity Gyroscope Experiment. The experiment has been developed by Francis Everitt following a suggestion by L. I. Schiff for a new test of the general theory of relativity. The test is based on the measurement of the precession of a gyroscope in orbit around the earth. The general theory predicts that the space–time curvature near the earth should cause the gyroscope to precess in two very specific ways. The first kind of precession is measured by the geodetic rate, Ω_G. It results from the motion of the gyroscope through the curved space–time around the earth. The other is the motional rate, Ω_M, resulting from the rotation of the earth. Both effects are tiny: the first should have a time-averaged value of 6.6 arcseconds per year; the second, .042 arcseconds per year. Even the attempt to measure them will produce new precessions that could totally swamp the effects the experiment is looking for. How is the effect of the space–time curvature to be isolated from all the other causal influences at work?

As we will see in the case of the Einstein–de Haas experiment to be described in section 3.1, the normal method for dealing with confounding effects is to calculate them from facts and theories already known, and then subtract them away. That is not what Everitt proposes to do. He intends not to calculate disturbing factors but to eliminate them.

With extreme care taken to minimize all other possible torques on the gyroscopes, *so that their resultant contribution to the drift rate is less than about 0.3 milli-arcseconds per year*, we expect a Gravity Probe-B Science Mission of 1 to 2 years to yield a determination of the relativity effects with a precision of better than 2 parts in 10,000 for Ω_G and better than 2 percent for Ω_M.[34]

The experiment was begun at Stanford by W. M. Fairbank and R. H. Cannon and it has been pursued since then by a joint team working with Everitt from the Stanford Physics and Aero-Astro Departments; Lockheed, Inc. is now working on the aerospace subcontractor; and there are hundreds of people involved in carrying out different phases of the development. Altogether it will have taken over twenty years to design this project from the time Everitt first started working on it until the gyroscope is eventually put into space in a NASA shuttle in 1991.

The precession will be measured with respect to the line of sight to a suitably chosen guide star, Rigel. We are told:

The concept for the Relativity Gyroscope Experiment is simple. The difficulty lies only in attaining the precision needed. Doing a 1 milli-arcsecond/year experiment requires a gyroscope with an absolute drift rate of about 10^{-11} degrees/hour (equivalent to about 5×10^{-17} radians/second or about 0.3 milli-arcseconds/year), some nine orders of magnitude better than current inertial navigation gyroscopes. In addition, the precision needed for line-of-sight determination to Rigel places severe design requirements on the telescope, its readout, and its attachment to the gyroscope reference structure.[35]

How do you get the drift rate from all the other sources to be less than 0.3 milli-arcseconds per year? (A football at a distance of 6,000 kilometres occupies 0.3 milli-arcseconds of the horizon.) Consider the problems just with the gyroscope itself. It must be almost perfectly spherical and perfectly homogeneous. It is calculated that the density gradient—that is, the departure from pure homogeneity—must be less than 3.6×10^{-7}. Commercially available fused quartz is almost that good, and that is what the rotor will be made from. The inhomogeneities in a material can be determined by looking for variations in the index of refraction. Simultaneously with the gyroscope experiment, a new instrument is being developed in the precision physics laboratory of M. Player in Aberdeen which uses sensitive interferometric techniques in a precisely controlled

---

[34]   D. Bardas *et al.*, 'Hardware Development for Gravity Probe-B', *Proceedings of SPIE* (International Society for Optical Engineering), vol. 619 (1986), p. 30.
[35]   Ibid.

temperature environment to measure variations in indices of refraction. Probably this instrument will be used to pick the material from which the final gyroscopes will be made.

A first major conceptual problem to be solved is how to measure the precession. After all, if you paint a spot on the ball and track it visually, that will destroy the homogeneity of the rotor. Instead the sphere will be coated with a very thin layer of superconducting material, which will create a magnetic moment aligned with the spin axis of the sphere when it rotates. Changes in the direction of the magnetic moment will be read out by a thin-film superconducting loop attached to a squid (*s*uperconducting *qu*antum *i*nterference *d*evice) magnetometer. The superconducting layer also allows the rotor to be suspended electrostatically. But it introduces a major disadvantage. For it means that the experiment must be conducted at cryogenic temperatures (~2K), and that produces its own sequence of problems. The need for cryogenic temperatures is over-determined, however; for the star-tracking telescope must also operate at liquid helium temperatures to remove distortion due to temperature gradients. The entire experimental module is shown in Fig. 2.9.[36] The dewar pictured there is designed to maintain cryogenic temperatures for one or two years in space.

Fig. 2.10 shows the fundamental requirements that the gravity probe experiment must satisfy if it is to pick out the effects of general relativity; and Fig. 2.11 shows a portion of the 'error budget' for those requirements. The total compounded error when all the systems are working together must be less than 0.3 milli-arcseconds per year. The error budget keeps track of all the separate sources and reckons how they will add. The simple philosophical point I want to make is most clearly illustrated in this budget: if we are going to bootstrap causes from the effects we can measure, we need to know how to account for all the other influences at work. Normally, it is impossible even to catalogue what these are, let alone calculate their effects. But that does not mean that we cannot engineer a situation where we can know what we need to know. The GP-B experiment is a good example. It may take twenty years, but in the end it should provide an entirely reliable test for the effects of space–time curvature, a test as stringent as any empiricist could demand.

I realize that this kind of reminder of the power of our practical

---

[36] I would like to thank Conrad Wiedemann for discussions of this example.

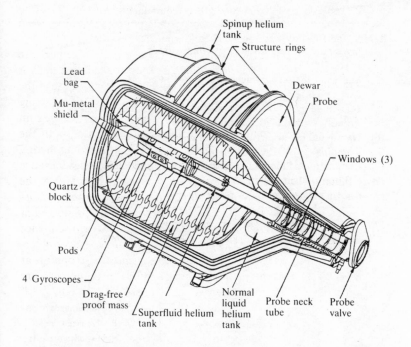

FIG. 2.9 *The GP-B experiment module showing the main elements of the dewar and probe*

Source: D. Bardas *et al.*, 'Hardware Development for Gravity Probe-B', *Proceedings of SPIE* (International Society for Optical Engineering), vol. 619 (1986), p. 30.

methods is no refutation of scepticism. But I am not trying to argue that knowledge in general is possible; only that causal claims as a class present no more difficulties than any others. If the GP-B experiment does not in the end prove to be a good enough test, that would not be because it is causes that we look for in our conclusions, nor because it is causal knowledge that we need as premises. Rather, it would be because we did not have enough of the knowledge that we needed; and that can happen in any experiment, whether we are trying to test for a causal process or to verify an equation. In both cases we frequently do know, or can come to know, what we need to know. We are secure in our conclusions because we are secure in our premises, and that security is no idle complaisance but a result of hard work and careful arrangement. When Everitt says his rotor is

Fundamental GP-B requirements

---

(1) Gyroscopes with < 1 msec/yr drift
(2) Gyro readout resolution better than 1 msec
    over : 100 sec
(3) Star tracker telescope accurate to < 1 msec
    over : 60 msec
(4) Integrated gyro housing telescope structure stable < 1 msec/yr
(5) Pointing within telescope linear range
(6) Data-handling accuracy better than 1 msec/yr with scale factor
    matching
(7) Elimination of long-term drifts from cyro and telescope readouts
(8) Calibration better than 1 msec
(9) Known proper motion of reference star

FIG. 2.10   *Fundamental GP-B requirements*
Source: L.S. Young, 'Systems Engineering for the Gravity Probe-B Program', *Proceedings of SPIE* (op. cit.), p. 55.

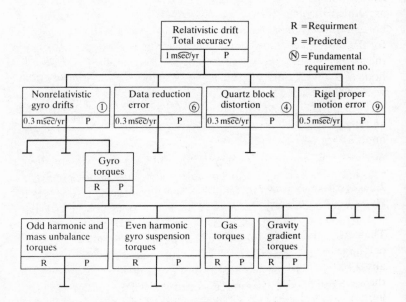

FIG. 2.11   *Sample of GP-B error tree*
Source: Young, op. cit., p. 56.

homogeneous to 3 parts in 10^{-7} this will be no mere wishful thinking on his part. Rather, he will have measured and found it to be so, using Player's new precession interferometric techniques.

We return in the end, of course, full circle to the original question. Grant that the information about the connection between space–time curvature and precession can be secured, at least as well as any other information. Must it necessarily be given an ineliminably causal interpretation? Can this seemingly causal claim not be read in some more Humean way? That is another question, indeed the question that takes up the rest of this chapter and much of the book. The special case of physics has already been discussed in section 2.2.

I argued at the beginning of this chapter that the necessity for background causal knowledge could not be avoided. The problem has on each occasion to be faced, and to be solved. I have given a rough sketch of the most dominant kinds of solution: try to construct an experiment that gets the conclusions indirectly without learning the causes one by one, or try to figure out exactly what they are and control for them. In practice both kinds of experiments borrow a little of both methods. Usually social scientists first control for the causes they know about, and then randomize; and in the end the gravity-probe experiment is going to roll the spacecraft in the hope of averaging out any causes they did not know about. Now I want to turn back to the discussion of why this knowledge is necessary, to pursue further the program of Glymour, Scheines, Kelly, and Spirtes, which promises to make this kind of knowledge unnecessary.

2.5. Discovering Causal Structure: Can the Hypothetico-Deductive Method Work?

The bulk of this book is directed against the Humean empiricist, the empiricist who thinks that one cannot find out about causes, only about associations. But it is a subsidiary thesis that causes are, nevertheless, very difficult to discover; and our knowledge, once we have left the safe areas that we command in our practical life, is not very secure, except in a very few of the abstract sciences. The kinds of method endorsed by Glymour, Scheines, Kelly, and Spirtes in *Discovering Causal Structure* would make causal inference easier—too easy for an empiricist, argued section 2.2. Yet for an adherent of causality, methods like theirs are probably the most natural alternative to the stringent requirement of measurement that I insist on. So it is important to look at more of the details to see how effective their kind of programme can be. The story I will tell is one of systematic and interwoven differences. We disagree, not just on the central issue of measurement versus the hypothetico-deductive method, but on a large number of the supporting theses as well. They tell one story and I another; and it probably speaks in favour of at least the consistency of both that the stories disagree through and through.

This section will lay out some of the details of these systematic differences. The reader who wishes to proceed with the main arguments of this chapter should go directly to the next section. A good deal of the hard work and the original ideas that have gone into *Discovering Causal Structure* are concerned with how to implement the basic philosophical point of view in a computer program called 'Tetrad'. I will have little to say about Tetrad, but instead will focus on the more abstract issues that are relevant to the theses I want to defend in this book.

The first and most important difference between my point of view and that argued in *Discovering Causal Structure* has already been registered. I insist that scientific hypotheses be tested. Glymour, Scheines, Kelly, and Spirtes despair of ever having enough knowledge to execute a reliable test; still, they think they can tell which structures are more likely to be true. To do so, they will judge theories by a combination of a simplicity requirement and Spearman's Principle. I do not hold with either of these requirements.

With respect to the first, the debate is an old and familiar one. They assume that structures that are simple are more likely to be true than ones that are complex. I maintain just the opposite. In *How the Laws of Physics Lie*[37] I have argued that nature is complex through and through: even at the level of fundamental theory, simplicity is gained only at the cost of misrepresentation. It is all the more so at the very concrete level at which the causal structures in question here are supposed to obtain. Matters are always likely to be more complicated than one thinks, rather than less. This view agrees exactly with that of Trygve Haavelmo in his early defence of the probability

---

[37] (Oxford: Clarendon Press, 1983).

approach in econometrics. Haavelmo argued that the most accept-able structures in economics will generally be more complex than at first seems necessary:

Every research worker in the field of economics has, probably, had the following experience: When we try to apply relations established by econo-mic theory to actually observed series for the variables involved, we fre-quently find that the theoretical relations are 'unnecessarily complicated'; we can do well with fewer variables than assumed a priori.[38]

What is the reason why it sometimes looks as if a simpler model will do? Haavelmo says that it is because, in some situations or over some periods of time, special conditions obtain. A large number of factors stay fixed, or average each other out. But these special conditions, he points out, cannot be relied on; so the original structures usually do not work when applied to different cases.

we also know that, when we try to make predictions by such simplified rela-tions for a new set of data, the relations often break down, i.e., there appears to be a *break in the structure* of the data. For the new set of data we might also find a simple relation, but a *different* one.[39]

Glymour, Scheines, Kelly, and Spirtes believe that simpler models are better. But I agree with Haavelmo. Simplicity is an artefact of too narrow a focus.

The issues involved in Spearman's Principle are less familiar, and the differing philosophical positions less well rehearsed; so it needs a more detailed discussion. A first thing to note about this principle is how relative its application is. For what is and is not a free parameter in a theory depends on what form the theory is supposed to take. Recall from Chapter 1 that there is a close connection between the linear equations of a causal model and the probability distribution of the variables in the model. It may help to begin by focusing on the probability distributions. It is usually assumed that the distributions involved will be multi-variate normal ones. This is the kind of constraint on the form of theory that I have in mind when I say that what counts as a free parameter in a theory depends on a prior specification of what form the theory must have.

For simplicity, consider just two variables, x and y, whose

[38] T. Haavelmo, 'The Probability Approach in Econometrics', *Econometrica*, 12 (1944), Supplement, 1–117.
[39] Ibid.

distribution is bi-variate normal. That means that the distribution in each variable considered separately (the marginal distribution) is normal; so too is the distribution for y in a cross-section at any point on the x curve, and vice versa; i.e. the conditional distribution $f(y/x)$ is normal and so is $f(x/y)$. Making the usual simplification that x and y both have mean 0 and variance 1, and representing their covariance by ρ, it follows by a standard theorem that the mean of the conditional distribution of y given a fixed value of x is ρx; that the conditional variance is $1 - \rho^2$; and that

$$y = \rho x + \mu$$

(where μ has mean 0 and variance $1 - \rho^2$).

Now consider the parameters, first for one variable, then for two. Given that a distribution in one variable is normal, the distribution is entirely specified by adding information about its mean and its variance; that is, the uni-variate normal distribution has just two free parameters. But this is not a necessary fact about a distribution. The exponential distribution has one; and for a totally arbitrary distribution one must specify not only the mean and the variance but an infinite sequence of higher-order moments as well. The point is a trivial one. The number of free parameters depends on how much information is antecedently specified about the distribution. In saying that the distribution in one variable is normal, one has already narrowed the choice from infinity to two; the fact that it is normal dictates what the higher-order moments will be once those two are fixed.

The bi-variate normal distribution has the five parameters described above: the mean and variance of each of the marginal distributions, plus the covariance of the two variables. But if one begins with the assumption that the distribution is a bi-variate normal with independent variables, then ρ must be equal to zero, and there are only four free parameters. Imagine, then, that one has (as in the example of cash-in-pocket and recidivism) data to show there is no correlation between x and y. Relative to one starting-point—the assumption that the distribution is bi-variate normal—the prediction of zero correlation is forthcoming only if one of the free parameters, ρ, takes on a certain specific value, in this case zero. Relative to the second starting-point the prediction follows, as in Model 1, no matter what values the free parameters take.

The same kind of point can be made by looking directly at the equations; and it is especially telling in the recidivism example. The most general set of consistent linear equations in *n*-variables with error terms included looks like this

$$(1) \quad a_{11}x_1 + \ldots + a_{1n}x_n = u_1$$
$$\ldots$$
$$(n) \quad a_{n1}x_1 + \ldots + a_nx_n = u_n$$

When these equations are supposed to represent causal structures, time-ordering can be used to bring the equations into a triangular array. The resulting form for a model in three variables is familiar from Chapter 1. Taking *r*, *u*, *c* as the three variables involved, it looks like this:

$$u = \text{exogenous}$$
$$c = \alpha u + w$$
$$r = \beta u + \gamma c + v$$

This is the general form for any causal model that involves just these variables essentially, and in the prescribed temporal order.

This is the form that is usually presupposed. But relative to this form, Spearman's Principle does not favour Model 1 over Model 2 in the recidivism example; rather, it does the opposite. Altogether there are three free parameters in the general form (excepting those that describe the distribution of the error terms): α, β and γ. In order to produce the prediction of zero correlation between *r* and *c*, Model 1 sets both $\beta = 0$ and $\gamma = 0$. Model 2 obtains the same result by setting $\beta = -\alpha\gamma$, where it is assumed that γ is negative and α and β are positive. In Model 1, just one parameter can still take arbitrary values; whereas in Model 2, two can. So Model 1 does not have the advantage by Spearman's Principle after all. Even if it did, it is not clear what that would signify. For the number of parameters necessary to account for the 'data' depends on what specific aspects of the data have been selected for attention. The correlation between *r* and *c* is just one aspect of the data. In principle one would want a model to account, not just for this or that correlation, but for the full probability distribution; and to do that, all the parameters would have to be fixed.

Glymour, Scheines, Kelly, and Spirtes avoid this problem by giving up the standard theory form. They proceed in a different order from what one might expect. They do not start with the usual sets of linear equations and then use Spearman's Principle to try to select among specific models. Rather, they use Spearman's Principle at the start to motivate a new general form for a causal theory. Here is how I would describe what they do. To get the new theory form, start with the old linear equations but replace all the usual continuous-valued parameters in the equations by parameters that take ony two values, zero and one. One can think of these new parameters as boxes, where the boxes are to be filled in with either a *yes* or a *no*; *yes* if the corresponding causal connection obtains and *no* if it does not. A specific theory consists in a determination of which boxes contain *yes* and which *no*.[40] This new theory form, they suppose, involves no free parameters (the two-valued ones do not seem to count). Here is where Spearman's Principle enters, to argue for the new theory form over the old.

The upshot of this implementation of Spearman's Principle is to reduce the information given in a causal theory from that implied by the full set of equations to just what is available from the corresponding causal pictures. The theory tells only about *causal structures*, that is, it tells qualitatively which features cause which others, with no information about the strengths of the influence involved. This move from the old theory form to the new one is total and irreversible in the Glymour, Scheines, Kelly, and Spirtes methodology, since the computer program they designed to rank causal theories chooses only among causal structures. It never looks at sets of equations, where numerical values need to be filled in. I think this is a mistake, for both tactical and philosophical reasons.

The philosophical reasons are the main theme of the remaining chapters of this book. The decision taken by Glymour, Scheines, Kelly, and Spirtes commits them to an unexpected view of causality. It makes sense to look exclusively at causal structures (i.e. the graphs) only if one assumes that (at least for the most part) any theory that implies the data from the causal structure alone is more likely to be true than one that uses the numbers as well. This makes causal laws fundamentally qualitative: it supposes that in nature only facts about what causes what are important; facts about strengths of influences are set by nature at best as an afterthought. I take it, by contrast, that the numbers matter, and that they can be relied on just as much as the presence or absence of the causal

---

[40] Assuming, as in the case of equations with ordinary parameters, that a choice of the variables has already been made.

relations themselves—and that that is a fact of vital practical significance.

The assumption that causal relations are stable lies behind the efforts of Glymour, Scheines, Kelly, and Spirtes, as it does behind the arguments of Chapter 1. In both cases, causal hypotheses are inferred from probabilities that obtain in one particular kind of situation, most preferably the situation of either a randomized or of a controlled experiment. But the point is not just to determine what happens in the special conditions of the experiment itself. The assumption is that the same causal relations that are found in the experiment will continue to obtain when the circumstances shift to more natural environments. But the same is true for hypotheses about strength of causal capacities. They too can be exported from the special circumstances in which they are measured to the larger world around. These claims are developed in later chapters. They enter here because they stand opposed to the view of Glymour, Scheines, Kelly, and Spirtes, which puts causal relations first. I think, by contrast, that much of nature is quantitative and that causal capacities and their strengths go hand-in-hand.

The tactical objections to restricting the admissible theories to just the causal structures are a consequence of problems that this decision raises for the concept of evidence. If there are no numbers in the theory, the theory will not be able to account for numbers in the data. One certainly will not be able, even in principle, to derive the full distribution from the theory. What, then, should the theory account for? Obviously one must look to the qualitative relations that hold in the data. Which ones?

Glymour, Scheines, Kelly, and Spirtes focus on correlations, and they pick two specific kinds of constraint on the relations among correlations that must be satisfied. The first requirement is that any relation of the form $\rho_{xz} - \rho_{xy}\rho_{yz} = 0$ must be satisfied, where ρ_{ab} represents the correlation between a and b. This factor is just the numerator of $\rho_{xz.y}$—i.e. the partial correlation between x and z controlling for y. Hence it is zero if and only if the partial correlation is zero. The partial correlation $\rho_{xz.y}$ is quite analogous to the conditional expectation $\text{Exp}(xz/y)$ that has been used throughout this book, and for purposes of the discussion here the reader can treat the two as identical. For Glymour, Scheines, Kelly, and Spirtes, then, the first constraint insists that the causal structure account for as many vanishing partial correlations in the data as possible.

The second constraint concerns tetrad relations, from which their computer program takes its name. These relations involve products of correlations among a set of four measured variables. Glymour, Scheines, Kelly, and Spirtes explain them this way: 'Tetrad equations say that one product of correlations (or covariances) equals another product of correlations (or covariances).'[41] For example,

$$\rho_{xy}\,\rho_{zw} = \rho_{xz}\,\rho_{wy}.$$

The choice that Glymour, Scheines, Kelly, and Spirtes make about which qualitative relations matter is very different from the one dictated by a methodology of testing. This is clear as soon as one reflects on how the relevant relations must be selected by an empiricist who wishes to use statistics to measure causes. In general the empiricist's rule is this: the data that are relevant are the data that will fix the truth or falsity of the hypothesis, given the other known facts. That means, in the context of causal modelling, that the relevant relations among the probabilities are those that tell whether the appropriate parameter is zero or not (or, if more precise information about the strength of the capacities is required, those that identify the numerical value of the parameter). Other relations that may hold, or fail to hold, do not matter.

Because Glymour, Scheines, Kelly, and Spirtes employ the hypothetico-deductive method, they must proceed in the opposite order. Their basic strategy for judging among models is two-staged: first list all the relevant relations that hold in the data, then scan the structures to see which accounts for the greatest number of these relations in the simplest way. That means that they need to find some specific set of relations that will be relevant for every model. But, from the empiricist point of view, no such thing exists.

Consider Glymour, Scheines, Kelly, and Spirtes, own choice. They ask the models to predict, from their structure alone, all and only correlations (that is, not the numerical values of these correlations but whether they exist or not) that hold in the data between two variables, with a third held fixed, and also any remaining tetrad relations that are not already implied by the vanishing partial correlations. This is an apt choice if all the variables have two causes, but not otherwise. With three causes, the relevant correlation between a putative cause and its effect is the one that shows up—or fails

[41] Glymour *et al.*, op. cit., p. 86.

to—with the other two held fixed; for four causes, three must be held fixed; and so forth; and when there is only one cause the correlation itself, with nothing held fixed, is what matters.

A better solution might be to include the higher-order (and lower-order) correlations as well: that is, to ask a structure to account for as many of the relations as *might* be relevant as it can. But that is bound to be wrong, since the very fact that makes an nth-order partial correlation the relevant one—that is, the fact that there are n other causes operating—also makes both higher- and lower-order ones irrelevant. Nor is there reason to think that some other choice will fare better. For what qualitative relations are relevant in the data depends on what causal structure is true; and each causal hypothesis must be judged against the data that are relevant for the structure they are, in fact, embedded in.

The difference in point of view about what data are relevant can be illustrated in the case that Glymour, Scheines, Kelly, and Spirtes themselves present to show what is wrong with methods, like those I have been defending, that try to infer causes from partial correlations. They discuss a study by Michael Timberlake and Kirk R. Williams that purports to show that 'foreign investment penetration [*f*] increases government repression [*r*] in non-core countries'. But, say Glymour, Scheines, Kelly, and Spirtes:

A straightforward embarrassment to the theory is that political exclusion is *negatively* correlated with foreign investment penetration, and foreign investment penetration is *positively* correlated with civil liberties and negatively correlated with government sanctions. Everything appears to be just the opposite of what the theory requires. The gravamen of the Timberlake and Williams argument is that these correlations are misleading, and when other appropriate variables are controlled for, the effects are reversed.[42]

The other variables Timberlake and Williams control for are energy development (*e*) and civil liberties (*c*). Controlling for these variables, they find that $\rho_{fr \cdot ec} \neq 0$; that is, foreign investment and repression are indeed correlated.

To keep the algebra simple, I will assume that the original (non-partial) correlation between foreign investment (*f*) and political repression (*r*) is not negative, but zero, which would be equally

---

[42] Ibid. 188.

damaging to Timberlake and Williams's hypothesis were no other causes at work.[43] We begin, then, by supposing that the data supports

$$d_0 : \rho_{fr} = 0$$

The model proposed by Timberlake and Williams then would be this:

Model TW:

$$f = \alpha g + u_1$$
$$c = \beta g + u_2$$
$$e = u_3$$
$$r = \delta f + \gamma c + \epsilon e + u_4$$

where it is assumed that γ is negative. Since it is assumed throughout that the initial nodes of a graph are uncorrelated, a new factor g has been introduced to produce the postulated correlation between f and c. The structure for Model TW is pictured in Fig. 2.12. The prediction that f and r are uncorrelated follows by setting $\alpha\delta = -\beta\gamma$.

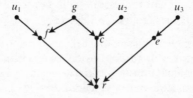

FIG. 2.12

Exactly the same thing is supposed to be happening in this structure as in the recidivism example: foreign investment causes political repression, but it is correlated with an equally powerful factor that prevents repression—civil liberties. Foreign investors tend to invest more in countries that already have a high level of civil liberties, and the two opposing factors exactly cancel each other, so that no positive correlation is observed between foreign investment and repres-

---

[43] I will also assume, for simplicity, that there is no correlation between f and e nor between c and e. Glymour *et al.* indicate that these correlations do exist; but they play no role in their discussion, so it seems reasonable to omit them.

sion, despite the causal connection between the two. Is this a likely story? Just as in the recidivism example, it takes a lot more information before that question can be answered. In particular, independent evidence is necessary for the auxiliary assumptions connecting energy development and absence of civil liberties with political oppression. Otherwise the appearance of a correlation between f and r, when e and c are held fixed, is entirely irrelevant. Apparently that is a problem in this study: '*Absolutely nothing* has been done to show that . . . [the] accompanying causal assumptions are correct.'[44] But that is a case of bad practice, and, as Glymour Scheines, Kelly, and Spirtes stress in defending their own methods, bad practices do not necessarily imply bad principles.

To focus on the underlying principles, then, for the sake of argument assume that the other two causal assumptions are reasonably well established (along with the assumption that these exhaust the causes); and, correlatively, that the data support the corresponding probabilistic relations

$$d_1 : \rho_{fr.ec} \neq 0$$
$$d_2 : \rho_{er.fc} \neq 0$$
$$d_3 : \rho_{cr.ef} \neq 0$$

These are the ones that matter for measuring the three causal influences.

Glymour, Scheines, Kelly, and Spirtes focus on different features from these three, since they look either for tetrad equations or vanishing three-variable partials. They find two relations in the data that are relevant under their criteria:

$$d_4 : \rho_{fr.e} = 0$$
$$d_5 : \rho_{ec.r} = 0$$

Looking at d_4 and d_5, and ignoring d_1, d_2, d_3, they favour a number of alternative structures, any one of which they take to be more likely than the one pictured in Fig. 2.12. Each of their structures reverses the causal order of r and c, making c depend on r, which is the easiest way to secure d_5. Since the methods described in Chapter 1 assume that temporal order between causes and effects is fixed, a structure in which f, c, and e all precede r, as they do in Model TW, will serve better for comparing the two approaches. Glymour, Scheines, Kelly,

---

[44] Glymour *et al.*, op. cit.

FIG. 2.13

and Spirtes doubt that foreign investment really causes political repression. Fig. 2.13 will do as an example of a structure that implies d_0, d_4, and d_5 just from its causal relations alone, keeps the original time-ordering, and builds in the hypothesis favoured by Glymour, Scheines, Kelly, and Spirtes that investment does not produce repression. Again the model includes an extra, unknown, cause, represented by g.

The example is an unfortunate one for Glymour, Scheines, Kelly, and Spirtes, however. For there is no way that this graph can account for the data d_1, d_2, d_3, with or without numbers. Nor is it possible with any other graph, so long as the time precedence of e, f, and c over r is maintained. If the original time-order is not to be violated, any model which accounts for d_4 and d_5 on the basis of its structure alone, and is consistent with d_1, d_2, and d_3 as well, must include the hypothesis that foreign investment causes repression.[45] This raises a problem of implementation. The Tetrad program designed by Glymour, Scheines, Kelly, and Spirtes will never find this out, since it never goes back to the full models that have numbers in them, and it never looks at any further relations in the data beside tetrads and three-variable partial correlations. Should the relevant data be extended to include the four-variable partial correlations supposed to hold in this case, then the structure pictured in Fig. 2.13 would be ranked very high, despite the fact that it is inconsistent with the data, since it can account for four of the five qualitative relations (d_3: $\rho_{cr.ef} \neq 0$ is the exception) on the basis of structures alone. One can, of course, check independently whether there is a corresponding

---

[45] This can be proved by considering what kinds of transformation could produce equations of the right form to imply the data. It also follows from Theorem (4.2) in *Discovering Causal Structure*, adding the observation that treks out of r can generate correlations with e only by introducing a new, ineliminable variable.

model consistent with all the data available before finally accepting a graph recommended by Tetrad. But in the all too common cases, like this one, where no such model exists, the Tetrad programme has no more help to offer.

For completeness, I give Model GSKS, which includes Glymour, Scheines, Kelly, and Spirtes' favoured hypothesis, that foreign investment does not cause repression, and which does account for all the data, though of course not on the basis of structure alone:

Model GSKS

$$f = \alpha c + \beta e + \gamma g$$
$$r = \delta c + \epsilon e + \phi g$$

where $\alpha\delta = -\gamma\phi$. The corresponding causal structure is given in Fig. 2.14.

Which is preferable of Model TW and Model GSKS? Model GSKS contains Glymour, Scheines, Kelly, and Spirtes' favoured hypothesis, that investment does not cause repression; Model TW says that it does. It is clear by now that an empiricist cannot resolve the question without more evidence. But neither can Glymour, Scheines, Kelly, and Spirtes. Even if one were willing to accept that simplicity, conjoined with Spearman's and Thurstone's Principles, was likely to take one closer to the truth, rather than further away, the three principles are of no help in this particular decision.

It is apparent that the decision about which kinds of data are relevant matters here. If tetrads and three-variable partial correlations were replaced by four-variable partial correlations, both Model TW and Model GSKS would fare very well, since d_1, d_2, and d_3 follow, in both cases, just from the graphs (assuming $\text{Exp}(g^2) \neq 0$). But the point of the examples is to make clear that that does not make the four-variable partials a better choice. Nor are they a worse choice. It

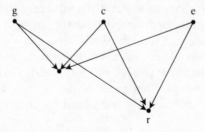

FIG. 2.14

is reasonable to demand that a model should account, from its structure alone, for just those relations in the data that hold because of the structure. But those relations cannot be identified in advance of any knowledge about what the structure is.

Given that they are, nevertheless, going to set out to choose some relations in advance, why do Glymour, Scheines, Kelly, and Spirtes opt for three-variable partials and tetrads? Practical problems of implementation provide a partial answer. But the choice is also based on principle. In the methodology of Chapter 1, three-variable partial correlations are significant because they identify causes in a three-variable model. That is not why Glymour, Scheines, Kelly, and Spirtes pick them. They pick them because of specific graph-theoretical results that show how three-variable partial correlations in the data generate constraints on the graph. In particular, their theorem (4.2) lays out some general conditions that any graph must satisfy if it is to account for vanishing correlations of this kind. Tetrad equations have a similar, though somewhat looser, motivation. Tetrad relations will occur when—though not only when—certain kinds of symmetry exist in the causal structure. What Glymour, Scheines, Kelly, and Spirtes themselves say about the vanishing three-variable correlations in the foreign investment example is this: 'These equations are interesting exactly *because they are the kind of relationship among correlations that can be explained by causal structure.*'[46]

This is an exciting attempt to secure universal relevance; but I do not think the connection is strong enough to do the job. The reason becomes clear by looking at a different kind of example, where problems of relevance arise regularly—say a scattering experiment. Consider an attempt to study the break-up of beryllium 9 by bombarding lithium 7 with deuterons. Two alpha particles and a neutron are produced. It is the alpha particles that will get detected. Since the final state involves three bodies, and not just two, the decay process may occur in a variety of different stages, so the model may be somewhat complicated. Imagine that the target is lithium sprayed on a carbon film, and that there are impurities in the target that produce irrelevant alpha peaks: the experimenters take measurements at a number of angles, and they find peaks at energies in the spectrum where they are not expected. In this case they do not use this data to build their model of beryllium 9; rather, they discard the

---

[46] Glymour *et al.*, op. cit., p. 190.

data because it is an artefact of the experimental set-up. Good methodology insists that they should have independent reasons in order to declare the data irrelevant. But the converse is what is at stake here: what does it take to make the peaks relevant? It is true that if the peaks *were* produced by the process under study, they *would* constrain the model in important ways. But that fact does not argue that they are relevant.

The same should be true for causal structures as well. Theorems like Glymour, Scheines, Kelly, and Spirtes (4.2) may seem to provide a way to mark out certain relations as relevant: the vanishing three-variable partial correlation is relevant to judging every structure because it implies some facts about what any structure will look like that can account for it. But the original question is always looming: *should* a structure be required to account for it? The model should do so, of course; but should it do so on the basis of structure alone? The point is almost a logical one. Glymour, Scheines, Kelly, and Spirtes take an implication that holds conditionally, and use it as if it were categorical. The fact that a particular kind of datum *would* dictate constraints on the structure of the phenomena under study if it *were* relevant does not bear—one way or the other—on whether it is relevant.

So the argument has come full circle. Glymour, Scheines, Kelly, and Spirtes do a masterful job tailoring the pieces—Spearman's Principle, Thurstone's Principle, and the demand for simplicity—to fit together into a coherent and, equally importantly, a usable whole. They themselves mention the possibility of expanding the program, to seek out higher-order correlations in the data and to look for structures that would account for these correlations; and they are always willing to balance criteria against one another. But, as an empiricist, I want neither more correlations, nor different ones. I want just those in each case that will measure the causes there.

2.6. Conclusion

It will be apparent that the conclusions reached so far in this chapter are somewhat at odds with the conclusions of the preceding chapter. So too is the methodology. This chapter develops a connection between probabilities and causes outside of any formal framework. It begins with the hypothesis that the introduction of a cause should increase the level of the effect; and then considers what further conditions must obtain to ensure that this increase will reveal itself in the probabilities. The result is formula *CC*, a formula that demands a lot in terms of background knowledge: what matters, according to this formula, is whether the cause increases the probability of the effect in populations where all other causally relevant features are held fixed. As section 2.4 argued, this does not necessarily mean that one has to know all the other causes in order to find out about any one of them. There are methods that circumvent the need for full information. Nevertheless, the justification for those methods rests on a formula that involves conditionalizing on a complete set of other causal factors.

Chapter 1 proceeded more formally; in the case of qualitative causes, by working with complete sets of inus conditions, and in the case of quantitative causes, by using sets of simultaneous linear equations. This means, in both cases, that very exact kinds of non-causal, or associational, information are presupposed: the truth of an equation, or the correctness of a set of conditions necessary and sufficient for the effect. It may seem that the use of these formal techniques avoids the need for conditioning on other causal factors. But that is not so. Before closing the chapter I want to point out exactly why. The reason lies in the open back-path-assumption. In order to draw causal conclusions from an equation, and thereby from the probabilities that identify the parameters of the equation, one must be assured that each factor in the equation has an open back path; and what that guarantees is that each of the factors is a genuine cause after all. So, in fact, the partial conditional probability that tells whether a particular parameter is zero or not—and thus tells whether a putative cause is genuine—is after all a probability that conditions on a complete set of other causal factors. The partial conditional probabilities that must be computed, either on the advice of Chapter 1 or on the advice of Chapter 2, are exactly the same. In a sense the open-back-path condition is doing the same kind of work that is done by the randomizations and controls of section 2.4: it side-steps the need for direct knowledge of the other causes by taking advantage of other knowledge that is more readily accessible.

There is a trade-off involved, however. For the open-back-path condition is quite cumbersome, and the accompanying proofs are bitty and inelegant. Both elegance and simplicity are restored if the condition is dropped and the devices of section 2.2 are resorted to:

just demand that the enterprise begins with a knowledge of what all the other causes are. This same device will work in the linear structures of Chapter 1 as well, and it is a device that was used by many of the early econometricians, whose methods are mimicked in Chapter 1. The experience in econometrics is worth reflecting on. It is apparent in the fifty-year history of that field that the appeal of this strategy will depend on how confident one is about getting the necessary starting knowledge. Theory is the natural place to look for it; and much of the early work in econometrics was carried through in a more optimistic mood about economic theory than is now common. When theory was in fashion, it was possible to adopt the pretence that the set of factors under consideration includes a full set of genuine causes. The problem then was to eliminate the phonies; and the methods of multiple correlation which were emerging in econometrics—and which are copied in simpler form in Chapter 1—are well suited to just that task.

This is one of the features that Keynes stressed in his criticisms of the new uses of statistical techniques advocated by Tinbergen. The remark quoted in Chapter 1 continues:

If we already know what the causes are, then (provided all the other conditions given below are satisfied) Prof. Tinbergen, given the statistical facts, claims to be able to attribute to the causes their proper quantitative importance . . .

Am I right in thinking that the method of multiple correlation analysis essentially depends on the economist having furnished, not merely a list of the significant causes, which is correct so far as it goes, but a *complete* list?[47]

As Keynes points out, the conventional methods are good for eliminating possibly relevant factors; but they are of no use unless the original set includes at least all the genuine causes. It is all right to start with too much, but not with too little. One can see this by returning to Fig. 1.5 (*b*). In that figure neither y nor x_1 appears, along with x_2, as a possible cause for x_e. Once both are included, the putative causes are no longer linearly independent (since $x_2 = ax_1 + y/g$). This means that the coefficients in the expanded equation, $x_e = ax_1 + bx_2 + cy + u_3$, are no longer identifiable and so cannot be determined from the data; the conventional methods will not tell which are zero and which are not.

[47] J. M. Keynes, 'Professor Tinbergen's Method', *Economic Journal*, 49 (1939), 560.

But linear independence bears only on identifiability, and it is clear from Chapter 1 that some additional argument is required to get from identifiability to causality. In this case the argument is easy—at least given something like Reichenbach's Principle, which is a necessary assumption in any attempt to forge a link between causes and probabilities. By hypothesis, the study is to begin with a set of variables $\{x_i\}$ which includes at least all the causes of the effect variable x_e, and possibly other factors as well.[48] The variables in this set must also satisfy a second condition: each is linearly independent of the others. Now an equation is given for x_e $(x_e = \Sigma a_i x_i)$. The equation is supposed to be true, but does it correctly represent the causal structure? If it does not, then by Reichenbach's Principle there must be another equation that does. By hypothesis this second equation will contain no new variables. But setting the two equations equal produces a linear relation among the members of $\{x_i\}$, which has been assumed impossible. So the original equation must give the true causal picture.

This is essentially the same line of argument as the one which guarantees identifiability; and, indeed, I am sure that this connection between the two ideas was intended by the early econometricians. Problems of identifiability, which were central to the work at the Cowles Commission, were for them problems of causality. This is discussed more fully in Chapter 4, but it is already apparent even in the piece by Herbert Simon described in Chapter 1. What Simon does is to solve a simple identification problem; what he says he does is to determine causal structure.

Why, then, are econometricians nowadays taught that identifiability and causality are distinct? One of the answers to this question illustrates the central thesis of this chapter: no causes in, no causes out. One has to start with causal concerns, and with causal information, to get causal answers. This is apparent in the comparison of Fig. 1.5 (*a*) and (*b*).

The equations of Fig. 1.5 (*b*) are identifiable. One may suppose, moreover, that they are true, and adequate for prediction. Yet Fig. 1.5 (*b*) does not picture the real causal structure, which is supposed to be given in Fig. 1.5 (*a*). The link between identifiability

---

[48] Or, more precisely, a set $\{x_i\}$ such that if x causes x_e then either x itself is in the set or x causes some factor that is.

and causality is missing because the equation for x_e associated with Fig. 1.5 (*b*) does not include all the causes.

These are typical of the kinds of equation produced nowadays in econometrics—either the structures are not causal at all, or they are mixed structures, where some of the variables are causal and others are not. This is apparent from the form of the structures—almost none has the recursive or triangular form that is induced by the time-ordering of causes and effects. Even those that do often add phenomenological constraints, by setting one quantity equal to another, without any pretence of representing the causal processes that produce the constraints.

Consider, for example, the continuation of Malinvaud's discussion, cited in Chapter 1, of non-recursive systems. Malinvaud introduces a classic model of the competitive market, a non-recursive, non-causal model, which he describes as interdependent. The model sets quantity demanded at any price equal to quantity supplied at that price. The process which leads to this equilibrium was described by L. Walras as a continuing negotiation between buyers and sellers. I quote at length from Malinvaud, to give a detailed sense of how the structure of this causal process comes to be omitted. Malinvaud points out that, in the usual econometric models,

effective supply, effective demand and price are finally determined at the same time. This is expressed by the interdependent system.

However, it is true that the model does not describe the actual process of reaching equilibrium, and therefore it obscures the elementary relationships of cause and effect. A different model would give a better description of the tentative movements towards equilibrium. . . .

[This example] shows why we must often make do with the interdependent models in econometric investigations. For it is impossible for us to observe the process of reaching equilibrium at every moment of time. We must be content to record from actual transactions the prices which hold and the quantities which are exchanged. Only the interdependent models . . . will interpret such statistics correctly.

More generally, available statistics often relate to relatively long periods, for example to periods of a year, in which time lags are much less significant. Thus, periods of a month would no doubt reveal a causal chain among income, consumption and production. But this model could not be used directly to interpret annual data. It is then better to keep to an interdependent system like the elementary Keynesian model, than to introduce the blatantly false assumption that consumption in year t depends solely on disposable income in year $t - 1$, which depends solely on production during year $t - 2$.[49]

There are other, more theoretical reasons for thinking that some given set of variables in an econometric model does not represent the basic causes at work: perhaps the whole level of analysis is wrong. But Malinvaud presents one simple set of reasons. Like any science whose basic methods depend on the estimation of probabilities, econometrics needs data. That means that it must use variables that can be measured, and variables for which statistics can be gathered. What kinds of model can be constructed with variables like that? Models that can be relied on for forecasting and planning? That is a deep question; a question that divides the realist from the instrumentalist. For the answer depends on how closely the power of a model to forecast and predict is tied to its ability to replicate the actual causal processes. What Malinvaud's remarks show is why the models actually in use are not causal, though they may well be identifiable. The reason supports the central thesis of this chapter: you can only expect your structures to be causal if you begin with causal variables.

[49] E. Malinvaud, *Statistical Methods of Econometrics* (Amsterdam: North-Holland, 1978), 57.

3

Singular Causes First

3.1. Introduction

How close can we come to a Humean account of causation? My
answer is, not very close at all. For Hume's picture is exactly upside
down. Hume began in the right place—with singular causes. But he
imagined that he could not see these in his experience of the world
around him. He looked for something besides contiguity to connect
the cause with the effect, and when he failed to find it, he left
singular causes behind and moved to the generic level. This move
constitutes the first thesis of the Hume programme: the only thing
singular about a singular causal fact is the space–time relationship
between the cause and its subsequent effect. Beyond that, the generic
fact is all there is. But the programme was far bolder than that, for at
the generic level causation is to disappear altogether. It is to be
replaced by mere regularity. That is the second thesis of the Hume
programme: generic causal claims are true merely by virtue of
regular associations.

Chapter 2 argued against the second of these theses: a regularity
account of any particular generic causal truth—such as 'Aspirins
relieve headaches'—must refer to other generic causal claims if the
right regularities are to be picked out. Hence no reduction of generic
causation to regularities is possible. This chapter will argue against
the first thesis: to pick out the right regularities at the generic level
requires not only other generic causal facts but singular facts as well.
So singular causal facts are not reducible to generic ones. There is at
best an inevitable mixing of the two levels.

I begin with a familiar puzzle. From Francis Bacon with his golden
events onwards, philosophers and scientists alike have acclaimed the
single instance. The classical probabilists like Bernoulli, Laplace,
and Poisson are a striking case, for they are the theorists who developed the mathematical theory of association; already by the end of
the eighteenth century their thinking had passed beyond Hume.

Although Hume was much indebted to the early classical probabilists for his account of probability, they were always wary of the value of associations as an aid to finding causes in science, and certainly none of them ever believed that associations constituted causation. In the best case, we need no recourse to associations (or probabilities) at all. Buffon cited the familiar example of the interlocking gears and weights of a clock, where the causes are manifest. Laplace worried that associations born of convention would create illusory subjective probabilities about causes. Poisson was willing to revamp the mathematical theory of induction to bring it into line with the actual practice of first-rate scientists. It does not take a lifetime of associations to convince a reasonable person of electromagnetic induction; Oersted's single experiment was quite sufficient.[1]

Poisson's point remains true in modern physics. The bulk of experiments that support the gigantic edifice of twentieth-century physics are never repeated, and they involve no statistics.[2] The trick of the outstanding experimenter is to set the arrangements just right so that the observed outcome means just what it is intended to mean; and that takes repeated efforts, usually over months and sometimes over years. But once the genuine effect is achieved, that is enough. The physicist need not go on running the experiment again and again to lay bare a regularity before our eyes. A single case, if it is the right case, will do.

This point is a familiar one. Nevertheless, it is worth looking at an example to make vivid the experience of working scientific investigation, against the abstract background of the metaphysical and epistemological props that support Hume's position. I choose as an example a series of experiments by Einstein and W.J. de Haas, experiments that led to the wrong conclusion but where, quite clearly, repeatability was not the issue. Einstein and de Haas went wrong, not because they tried to establish a general truth from what they saw in the single case, but rather because they mis-identified what they actually saw.

In 1914 Einstein and de Haas set out to test the hypothesis that magnetism is caused by orbiting electrons. They tested it by suspend-

---

[1] L. Daston, *Classical Probability in the Enlightenment* (Princeton, NJ: Princeton University Press, 1988).

[2] Though see P. Galison, *How Experiments End* (Chicago, Ill.: Chicago University Press, 1987).

ing an iron bar in an oscillating magnetic field and measuring the gyrations induced when the bar was magnetized. They expected the bar to oscillate when the field was turned on and off because electrons have mass, and when they start to rotate they will produce an angular momentum. The ratio of this momentum to the magnetic moment—called the gyromagnetic ratio—should be $2m/e$, where m and e are the mass and charge of the electron respectively. This is very close to the answer Einstein and de Haas found. But it is not the result they should have got. Later experiments finally settled on a gyromagnetic ratio about half that size, and nowadays—following the Dirac theory—the results are attributed, not to the orbiting electrons, but to a complex interaction of orbit and spin-orbit effects.

What went wrong with the Einstein–de Haas experiment? The answer is—a large number of things. I take this example from a paper by Peter Galison;[3] the reader can see the complete details laid out there. Galison describes the work of ten different experimental groups producing dozens of different experimental constructions over a period of ten years to establish finally that the Einstein–de Haas hypothesis was mistaken. I will briefly discuss just one of the factors Galison describes.

Besides the effects of the hypothesized electron motion, it was clear that the magnetic field of the earth itself can also cause a rotation in the bar, so there had to be a shield against this field. 'At first [Einstein and de Haas] used hoops with a radius of one meter with coils wound around them to eliminate the earth's field.'[4] In the next set of experiments de Haas wrapped the wire of the solenoid as well. He also arranged a compensating magnet near the centre of the bar, and two near the poles, as well as a neutralizing coil at right angles to the bar. In 1915 Samuel Barnett from Ohio State University performed similar experiments with a great number of improvements. In particular, he neutralized the earth's field with several large coils. As Galison reports, 'the outcome after his exhaustive preparations was a value [of the gyromagnetic ratio] less than half of that expected for orbiting electrons.'[5] The story goes on, but this is

---

[3] 'Theoretical Predispositions in Experimental Physics: Einstein and the Gyromagnetic Experiments, 1915–1925', *Historical Studies in the Physical Sciences*, 12(2) (1982), 285–323.

[4] Ibid. 298.

[5] Ibid. 320.

enough to give a sense of the detail of thought and workmanship necessary to get the experiment right. The point is that we sometimes *do* get it right, and when we do, we can see the individual process that we are looking for, just as we see the process in Buffon's clock; and that is enough to tell whether the causal law is true or not.

The puzzle I raise about the lack of fit between the practical reliance on the single case versus the philosophical insistence on the primacy of the regularity is in no way new. Yet familiarity should not make us content with it; Hume's own view that we can lay our philosophy aside when we leave the study and enter the laboratory is ultimately unsatisfactory. In fact, he fatally failed to distinguish the laboratory from the rest of the world outside the study. Both the philosopher on one side and the experimentalist on the other must be concerned when epistemology and methodology diverge.

I realize that there are a number of ways in which the Humean can try to account for the schism between philosophy and practice. Admittedly, one-shot experiments, like those of Oersted, Einstein and de Haas, or Barnett, work in disciplines like physics where there is a gigantic amount of background information, precise enough to guarantee that the experiment isolates just the one sequence of events in question. The logic of these experiments involves a complex network of deductions from premises antecedently accepted, and a good number of these premises are already causal. Perhaps it is not surprising from the Humean point of view that singular confirmation is possible once one is operating within such a large set of assumptions.

But it must be surprising that no causal conclusions are possible outside such assumptions. Without antecedent information it is no more possible to establish a causal claim via a regularity than it is to demonstrate a singular cause directly; and in both cases the inputs must include causal information—not only information about general causal laws, but about singular facts as well. This is the argument on which I will concentrate in this chapter, because it attacks the regularity view directly. Arbitrary regularities do not amount to causal connections. Which regularities do? My basic claim is that figuring that out is a laborious job that must be undertaken anew in each new case, as the kinds of known causal structure in the background differ. In this chapter I will consider a few simple and idealized examples just to show that, in any case, information about singular causes is vital.

The chapter ends with a more radical doctrine. Singular claims are not just input for inferring causal laws; they are the output as well. At the beginning of the introduction I said that the chapter would show how the generic and the singular are inextricably intertwined. But the ultimate conclusion is far stronger. Singular facts are not reducible to generic ones, but exactly the opposite: singular causal facts are basic. A generic claim, such as 'Aspirins relieve headaches', is best seen as a modalized singular claim: 'An aspirin can relieve a headache'; and the surest sign that an aspirin can do so is that sometimes one does do so. Hence my claim that Hume had it just upside down.

3.2. Where Singular Causes Enter

If the last chapter is correct, probabilities by themselves can say nothing about the truth of a general causal hypothesis. A good deal of information about other causal laws is needed as well. But that does not exhaust the information required: not only must other general causal claims be supposed, but information about singular causal facts must be assumed as well. To see why, return to formula CC of Chapter 2. Formula CC says that, for a generic causal claim to hold, the putative cause C must increase the probability of the effect E in every population that is homogeneous with respect to E's other causes. But this condition is too strong, for it holds fixed too much. The other factors relevant to E should be held fixed only in individuals for whom they are not caused by C itself. The simplest examples have the structure of Fig. 3.1.

This is a case of a genuine cause C, which always operates through some intermediate cause, F. But F can also occur on its own, and if it does so, it is still positively relevant for E. Holding F fixed leads to the mistaken conclusion that C does not cause E. For

FIG. 3.1

$P(E/C \pm F) = P(\neg C \pm F)$. This is a familiar point: intermediate causes in a process (here F) screen off the initial cause (C) from the final outcome (E). If intermediates are held fixed, causes will not be identified as genuine even when they are. On the other hand, if factors like F are not held fixed when they occur for independent reasons, the opposite problem arises, and mere correlates may get counted as causes.

What is needed is a more complex characterization of the precise way in which a population must be homogeneous. *CC* must be amended to ensure that

* Each test population of individuals for the law 'C causes E' must be homogeneous with respect to some complete set of E's causes (other than C). However, some individuals may have been causally influenced and altered by C itself; just these individuals should be reassigned to populations according to the value they would have had in the absence of C's influence.

This means that what counts as the right populations in which to test causal laws by probabilities will depend not only on what other causal laws are true, but on what singular causal processes obtain as well. One must know, in each individual where F occurs, whether its occurrence was produced by C, or whether it came about in some other way. Otherwise the probabilities do not say anything, one way or the other, about the hypothesis in question.

A very simple and concrete example with the problematic structure pictured above has been given by Ellery Eells and Elliot Sober.[6] I expand it somewhat to illustrate how both holding F fixed and failing to hold it fixed can equally lead to trouble. Your dialling me (C), they suppose, causes my phone to ring (F), and my phone's ringing causes me to lift the receiver (E). 'So presumably your phoning me thus causes me to lift the receiver.'[7] But this claim will not be supported by the probabilities if the ringing is held fixed, since $P(E/C \pm \text{F}) = P(E/ \neg C \pm F)$; that is, once it is given that the phone rings, additional information about how it came to ring will make no difference. To require the contrary 'would mean that your calling me at t_1 must have a way of affecting the probability of my picking up the phone at t_3 other than simply by producing the ringing

---

[6] E. Eells and E. Sober, 'Probabilistic Causality and the Question of Transitivity', *Philosophy of Science*, 50 (1983), 35–57.

[7] Ibid. 40.

at t_2'.[8] Holding fixed F in this case would give a misleading causal picture.

On the other hand, not holding F fixed can be equally misleading, for reasons which are by now familiar. Imagine that you phone me in California every Monday from the east coast as soon as the phone rates go down. But on each Monday afternoon another friend, just a little closer, does the same at the same time, and you never succeed in getting through. In this case it is not your phoning that causes me to lift the receiver; though that may look to be the case from the probabilities, since now $P(E/C) > P(E/\neg C)$. But the causes and the probabilities do line up properly when F is held fixed in the way recommended by Principle *. Consider first the $\neg F$ population. This population should include all the Monday afternoons on which my phone would not otherwise ring. On these afternoons your dialling does cause me to lift the receiver, and that is reflected in the fact that (given $\neg F$) $P(E/C) > P(E/\neg C)$ in this population. In the second population, of afternoons when my phone rings but because my other friend has called, your dialling does not cause me to lift the receiver, nor is that indicated by the probabilities, since here (given F) $P(E/C) = P(E/\neg C)$.

There are a number of ways in which one might try to avoid the intrusion of singular causes into a methodology aimed at establishing causal laws. I will discuss four promising attempts, to show why they do not succeed in eliminating the need for singular causes. The first tries to side-step the problem by looking at nothing that occurs after the cause; the second considers only probabilities collected in randomized experiments; the third holds fixed some node on each path that connects the cause and the effect; and the fourth works by chopping time into discrete chunks. The first two suffer from a common defect: in the end they are capable of determining only the 'net upshot' of the operation of a cause across a population, and will not pick out the separate laws by which the cause produces this result; the third falters when causes act under constraints; and the fourth fails because time, at least at the order of magnitude relevant in these problems, does not come already divided into chunks. The division must be imposed by the model, and how small a division is appropriate will depend on what singular processes actually occur. I will begin by discussing strategies (i) and (ii) in this section, then

---

[8] Ibid.

interrupt the analysis to develop some formalism in section 3.3. I return to strategies (iii) and (iv) in sections 3.4.1 and 3.4.2.

3.2.1. Strategy (i)

The first strategy is immediately suggested by the telephone example; and indeed it is a strategy endorsed by the authors of that example. Principle *, applied in this example, describes two rather complicated populations: in the first, every Monday afternoon which is included must be one in which my phone rings, but the cause of the ringing is something different from your dialling. By hypothesis, the only way your dialling can cause me to lift the receiver will be by causing my phone to ring. So in this population it never happens that your dialling causes me to lift the receiver; and that is reflected in the probabilities, since in this first population $P(E/C) = P(E/\neg C)$. But matters are different in the second population. That population includes the Monday afternoons on which my phone does not ring at all, and also those ones on which it does ring, but the ringing is caused by your dialling. In this population, your dialling does cause me to lift the receiver; and, moreover, that is also the conclusion dictated by the probabilities, since in the second population $P(E/C) > P(E/\neg C)$.

But in this case there is an altogether simpler way to get the same results. The ringing never needs to come into consideration; just hold fixed the dialling of the second friend. When the other friend does dial, since by hypothesis she gets connected first, your dialling plays no causal role in my answering the phone; nor does it increase the probability. When she does not dial, you do cause me to lift the phone, and that is reflected in an increase in the probability of my doing so. This is just the strategy that Eells and Sober propose to follow in general. Their rule for testing the law '*C* causes *E*' is to hold fixed all factors prior to or simultaneous with *C* which either themselves directly cause (or prevent) *E*, or which can initiate a chain of factors which can cause (or prevent) *E*. Their picture looks like Fig. 3.2.[9] It is not necessary to hold fixed causes of *E* that occur after *C*, argue Eells and Sober; holding fixed all the *causes* of these causes will succeed in 'paying them their due'.[10]

<hr>

[9] Ibid. 40–1. Also personal correspondence.

[10] More formally, to test the causal law $C_t' \rightarrow E_t''$, Eells and Sober hold fixed all factors K_t *such that* $t > t'$ and (i) $K_t \rightarrow \pm Et''$ or (ii) there exists a chain $F_1(t_1), \ldots, F_n(t_n)$ such that $K_t \rightarrow F_1(t_1) \rightarrow F_2(t_2) \rightarrow \ldots \rightarrow \pm E_t''$, for $t < t_1 < \ldots < t_n < t''$. (Here $C \rightarrow E$ means '*C* causes *E*.')

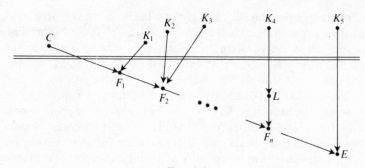

Fig. 3.2

This is true in simple cases where causes operate in only one way to produce or prevent the effect. But often a factor has mixed capacities—it can both cause and prevent the same effect, or cause it in different ways with influences of different strengths. An example common in the philosophical literature comes from G. Hesslow.[11] Hesslow argues that birth-control pills both inhibit and encourage thrombosis. Let C represent the contraceptives; T, thrombosis. He advocates then that not one but two causal laws are true: 'C causes T' and 'C prevents T'. The pills prevent thrombosis by preventing pregnancy (P), which itself tends to produce thrombosis. On the other hand, they themselves frequently cause thrombosis. Hesslow does not specify any intermediate steps in the positive process, but one can imagine that the pills produce a certain chemical, C', that causes the blood to clot and thereby produces thrombosis. Hesslow's hypotheses are represented in Fig. 3.3. Fig. 3.3 follows the usual conventions and identities 'A prevents B' with 'A causes $\neg B$'.

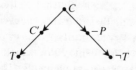

Fig. 3.3

For simplicity, imagine that pregnancy and the chemical C' are the only factors relevant at t_2 for producing or inhibiting thrombosis at t_3. It will help to keep the structure as simple as possible by

[11] 'Discussion: Two Notes on the Probabilistic Approach to Causality', *Philosophy of Science*, 43 (1976), 290–2.

assuming that the only way a factor at t_1 bears on thrombosis at t_3 is either via pregnancy or via C'; and also to treat all those factors that are relevant at t_1, other than the contraceptives themselves, together as a single general background which will be labelled B. In this case the Eells–Sober strategy for judging the effects of pills on thrombosis is to hold fixed B; and this is a sensible strategy from the point of view of the problems raised so far. For the familiar problems of joint effects, and of other related kinds of 'spurious correlations', arise when there are background correlations for some reason or another between the putative cause and other causal factors. In this case, by construction there are only two independent factors with which C might be correlated—C' and P. But if all other causes of C' and P are held fixed, there is no way for the contraceptives to be correlated with these, other than by their own causal actions. This is what Eells and Sober mean by 'paying them their due'.

Unfortunately, background correlations are not the only source of problems. The dual capacity of the contraceptives also makes trouble. Because the contraceptives can act in two different, opposed ways, their probabilistic behaviour will be different in different circumstances: in one kind of circumstance they push the probabilities up; in another they push them down. If these different circumstances are not kept distinct; but instead are lumped together, these opposing probabilistic tendencies can get averaged out, so that at best one, but possibly neither, of the opposing capacities will be revealed. This is easy to see in the four kinds of causally homogeneous populations produced by B: (1) $C'P$, (2) $C' \neg P$, (3) $\neg C'P$, and (4) $\neg C' \neg P$. The first population is one in which every woman is both pregnant and has the chemical C' in her blood; in the second, no one is pregnant, but all have the chemical; and so forth. In the absence of C, it can be supposed that B produces these four populations in some fixed ratio, and the resulting level of thrombosis in the total group will be an average, with fixed weights, over its level in each of the four homogeneous populations taken separately.

What happens if B does not act on its own, but C occurs as well at t_1? If the contraceptives do indeed affect both pregnancy and the amount of chemical in the blood, the ratios among these four populations will change.[12] The second group, of women who have C' at t_2 and are not pregnant then, will stay at least as big as it was;

----

[12] This assumes that the capacity of C to affect P and to affect C' remains the same in the presence and in the absence of B. Cf. ch. 5.

since the contraceptives cause C' and prevent pregnancy, they will not change the situation of anyone who would otherwise have had C', or who would not have been pregnant in any case. Indeed, this group will grow larger; for it will receive additions from all the other groups. In the first group, the contraceptives will have no effect on the rate of C', but they will prevent some pregnancies which would otherwise have occurred. Thus, some women who would have been in Group 1 under the action of B alone will move into Group 2 when C acts as well. Similar shifting occurs among the other groups. The group that has both effects already must necessarily grow bigger; and the group with neither effect will in the end be smaller; what happens in the other two depends on whether the tendency of the contraceptives to induce C' is stronger or weaker than its tendency to inhibit pregnancy.

The net result for thrombosis of all these changes is unpredictable without the numbers. It depends not only on how effective C is, versus B, in producing the harmful chemical and preventing pregnancy, but also on how effective the chemical and pregnancy themselves are in producing thrombosis. Anything can happen to the overall probability. If the processes that operate through the prevention of pregnancy dominate, the number of cases of thrombosis will go down when contraceptives are taken; conversely, if the processes operating through the chemical dominate, the number will go up; and in cases where the two processes offset each other, the number will stay the same. But this does not in any way indicate that contraceptives have no power to cause or to prevent thrombosis, any more than the dominance of their good effects would show that they had no negative influence, or vice versa. No matter how the relative frequencies work out, the pills are both to be praised and blamed. In any case, they will have caused a number of women to get thrombosis who would otherwise have been healthy; and this fact is in no way diminished by the equally evident fact that they also prevent thrombosis in a number of women who would otherwise have suffered it. It is true that in either case the effect is achieved through some intermediary. The pills cause thrombosis by causing C' where it would not otherwise occur; similarly, they prevent thrombosis by preventing pregnancies that would have occurred. But that is hardly an argument against their power. Since, at least at the macroscopic level, causal processes seem to be continuous, all causes achieve their effects only through intermediaries.

The lesson to be learned from this case is that the strategy urged by Eells and Sober to avoid the mention of singular causes will not work when causes have mixed capacities. But will Principle *, which does rely on information about the single case, fare better? It should be apparent that the answer is yes. For this proposal involving singular causes retraces the argument that was just made. It says: to uncover the connection between contraceptives and thrombosis, assign individuals to groups on the basis of whether they would have C' and P if C did not operate. Then consider, in each of these groups separately, how frequent thrombosis is among women who take contraceptives versus its frequency among women who do not. What the earlier argument showed is that both the positive capacity and the negative capacity of the contraceptives are bound to come out in this procedure, since in Group 4 the incidence of thrombosis will surely go up and in Group 1 it will surely go down.

This is, moreover, exactly the strategy that anyone would advocate, including Eells and Sober themselves, were it not for the awkward question of timing. Imagine, for instance, a slightly altered example in which B operates exactly as before in producing C' and P, but in which it operates a little earlier than C, so that C' and P are already in place before the pills are taken.[13] In this case the need for the inconvenient singular counterfactual completely disappears. The action of the contraceptives moves women, not from groups they would have been in, but from groups they are in. By stopping pregnancies that would otherwise occur under the action of B alone, they prevent thrombosis; and by producing the chemical C when it did not exist before, they cause thrombosis. The results are apparent in the frequency of thrombosis in Group 1 ($C'P$) where the probability will be less with C than without; and in Group 4 ($\neg C' \neg P$), where the converse holds. In Group 2 ($C' \neg P$), C can have no effect, and in Group 3 ($\neg C'P$) the effects are mixed.

It is obvious in the case of the altered example that the four groups must be kept separate: C' and P should be held fixed. When they are not, the consequent probability of the effect will be an average over its probability in each of the four groups separately; and when four

---

[13] To make sense of this in the case of pregnancy, one must imagine that B creates some kind of a fixed capacity guaranteeing that the individual will either surely get pregnant, or surely not, unless some further factor intervenes. This is obviously a made-up assumption, but it is worth stretching the plausibility of the example to keep the structure of the argument as clear as possible.

different outcomes pointing in different directions are averaged, anything may result. Conventional wisdom teaches that averaging must be avoided in this case. Yet it is exactly this same averaging—with its untoward consequence—that results from the Eells and Sober strategy. Holding fixed only the factors that occur up to the time of the cause produces a population that mixes together the various different groups which need to be considered separately. It makes no difference whether the independently occurring causes take place before C—as in the original example—or after C—as in the altered example. They must not be averaged over in any case.[14]

3.2.2. Strategy (ii)

The second strategy for eliminating the need for singular causes is to look only at the probabilities from randomized experiments and to see whether there is a higher frequency of the effect in the treatment group, where the cause has been introduced, than in the control group, where it has been withheld. Recall from Chapter 2 that randomized experiments go a long way toward eliminating the need for background causal knowledge. In particular, since the treatment is supposed to be introduced independently of any of the processes that normally occur, problems of spurious correlation can never arise. But the probabilities that show up in a randomized experiment, even in a model experiment where all the ideal specifications are met, will not reveal the true capacities which a cause may have. For conventional randomized experiments average over subsequently occurring causes in the way that has just been illustrated.

Consider the case of the birth-control pills. The standard randomization procedures are supposed to guarantee that the distribution of various arrangements of the background factors, summarized in B, will be identical in the treatment and the control group. This in turn should ensure that the relative frequencies of each of the effects of B are the same in both groups. But obviously neither the test nor the control group will be homogeneous with respect to these effects. Conceptually, each group could be segmented into the four sub-populations of the previous discussion, each homogeneous with respect to B's effects. But the separation is

[14] For a detailed proof that the two averagings are indeed exactly the same, see N. Cartwright, 'Regular Associations and Singular Causes', in B. Skyrms and W. L. Harper (eds.), *Causation, Change, and Credence*, i (Dordrecht: Reidel, 1988), 79–97.

not made in the experiment; and the final probability for the effect inevitably averages over the probabilities in each of these four separate populations. What the experiment reveals is the net result of the operation of the cause across a population, disentangled from any confounding factors with which that cause might normally be correlated. This kind of information is extremely useful for social planning, and possibly even for personal decision-making. But it does not exhaust the causal structure. As has already been stressed, a cause whose net result across the population is entirely nil may nevertheless have made a profound difference, both in producing the effect where it would not otherwise have been and in preventing it where it otherwise might have been.

There remain two further strategies to be discussed. But it will help in proceeding in an orderly manner to balance the kind of intuitive argument I have been using so far, based on seeing what is at stake in various kinds of hypothetical example, with a tidier kind of argument that depends on a more formal structure. So the discussion of these remaining strategies will be delayed until section 3.4; before that, in section 3.3, a formal apparatus will be developed that will bring some system into my discussion of probabilistic causality.

3.3. When Causes Are Probabilistic

This section will show how to modify the conventional equations of a linear causal model to incorporate causes which act probabilistically. It will consist of three parts.

The first part will explain what notion of probabilistic causality is intended, and will show why the causes that are represented in the conventional linear equations are not probabilistic, despite the appearance of random error terms in those equations; and this section will finally suggest a simple way to amend the equations to make the causes probabilistic.

The second part uses the modified equations to get a clearer picture of what assumptions about causal structure are built into the original formalism. Using the new notation, it is easy to see that the conventional assumptions about the independence of the error terms in the standard equations presuppose that all causal processes operate independently of all others. This means that the standard representation has a quite restricted domain of application. A simple three-variable example will be given to show how much difference the independence assumption makes. The example involves a cause which produces two different effects, but subject to a conservation principle. As an illustration, at the end of the second part I will show how the familiar factorizability criterion for a common cause fails in this case, and what must be put in its place.

The third part looks ahead to see how this can make a difference to questions about causality in quantum mechanics.

3.3.1. A New Representation

In the usual equations of a causal model, the functional relation between a cause and its effect is exact. Whatever value the cause takes, it is bound to contribute its fixed portion to the total outcome. But often the operation of a cause is chancy: the cause occurs but the appropriate effect does not always follow, and sometimes there is no further feature that makes the difference. In the terminology of G.E.M. Anscombe,[15] the cause is *enough* to produce the effect, though it need not be sufficient to guarantee it.

It is possible for a chancy cause to operate entirely haphazardly. Sometimes it produces its effect and sometimes it does not, and there is no particular pattern or regularity to its doing so. I am going to ignore these cases and focus instead on causes that are better behaved—on purely probabilistic causes, causes which, when in place, operate with a fixed probability. Radioactive decay is a familiar example. A uranium nucleus may produce an alpha particle in the next second, and it may not; but the probability that it will do so is an enduring characteristic of the nucleus. Obviously more complicated cases are imaginable, cases in which not only is it a matter of chance whether the cause contributes its influence or not, but where the degree, or even the form, of the influence is only probabilistically fixed. I shall deal only with the simpler cases, where the form of the influence is fixed and only its occurrence is left to chance.

Before considering how best to deal with probabilistic causes, a short digression on adding influences will probably be of help. Econometricians sometimes say that the equations they study need not be linear in the variables, but only in the parameters. This means

---

[15] *Causality and Determination: An Inaugural Lecture* (London: Cambridge University Press, 1971).

that the relation between a cause and the influence it contributes need not be linear at all. In the case of gravitational attraction, for example, the distance r and the masses m_1 and m_2 are partial causes which together contribute an influence of the form Gm_1m_2/r^2. What is important is that the respective influences are additive. This is a point that the economist Tinbergen made explicit early on, in reply to Keynes's objection that the various factors which get added together do not have the same units: 'I do not add up the "factors" (in my terminology the "explanatory variables"), but I add up their "influence". . . .'[16]

To assume, as Tinbergen says, that the influences are additive is to assume that the separate causes do not interact. This is a topic that will be taken up in Chapter 4. Here I want to stress a point that helps to explain why regressions and correlations are of such limited use in physics. The correlational methods associated with causal models begin with variables which are presumed to add. Since it is only the influences and not the causes themselves that can reasonably be expected to be additive, these methods begin to apply only after the influences have been settled on. The methods of econometrics, and indeed of most probabilistic studies of causality, are of little help in determining the form of the influence that a cause contributes; rather, they are designed to find out whether the cause really contributes at all, given that the form of its contribution is assumed. This is a fact often concealed by the notation. Assuming that the influences are additive, it would be more perspicuous to write the effect variable (x_e) as a function of its causes like this:

$$x_e = a_1 f_1 (x_1') + a_2 f_2 (x_2') + \ldots + a_n f_n (x_n') + u_e$$

Using the relation $x_n = f_n(x_n')$ gives the more familiar-looking equations of Chapter 1. I shall keep to this standard notation, and thus I shall talk as if the cause is x_n, and its influence $a_n x_n$. But in fact x_n would be better thought of in most cases as itself already an influence of something else that would more naturally be called the cause.

How should probabilistic causes be incorporated into this traditional scheme? At first sight it may seem that they are already there in the us. These are, after all, conventionally called 'error terms'.

---

[16] J. Tinbergen, 'On a Method of Statistical Business-Cycle Research: A Reply' *Economic Journal*, 50 (1940), n. 197, pp. 141–54, 147. Though note that what in fact Tinbergen adds is 'the product of the variable and its regression coefficient', and hence his exogenous variables are already influences, as suggested here in the text.

Will that not by itself turn a scheme that looks deterministic into one that is stochastic? It is a central thesis of this section that the answer to this question is no. A liberal interpretation of the 'errors' can indeed introduce a random element; but it will not easily turn deterministic causes into probabilistic ones. To see why, consider how these terms have been traditionally interpreted.

Historically, error terms did not formally appear in equations until the 1940s.[17] Since then they have been variously interpreted as representing omitted variables, random shocks, non-economic forces, individual differences, and more. But the point about probabilistic causes is easy to grasp by considering either of the two most usual interpretations: the 'errors-in-variables' interpretation and the 'errors-in-equations' interpretation. The errors-in-variables interpretation takes the extra terms in the equations to represent measurement error, which occurs in trying to observe the true cause. The aim is to find the relationship between the effect x_e and its true causes x_1, ..., x_n, given information about the observed values x_e', x_1', ..., x_n'. In general the observed values do not match the true ones; each observation may include some error: $x_1' = x_1 + u_1$, ..., $x_n' = x_n + u_n$, so $x_e = x_1' + \ldots + x_n' + u$, where $u = u_1 + \ldots + u_n$. Clearly here the causes are not probabilistic. The empirical assessment of the influence may be mistaken, but each cause is bound to contribute its full influence on each occasion.

This is equally true of the second reading, where u is supposed to represent the sum of the contributions from all the causes that have not been explicitly mentioned, i.e. causes other than those represented by $x_1, \ldots, x_n$. It is called the 'errors-in-equations' interpretation because the equation in the xs alone would literally be in error, unless something is added to represent the missing factors. But what is added is itself another cause, and each cause, whether represented by a u or by an x, operates entirely deterministically: x_n always contributes exactly $a_n x_n$ to the effect. There is never any possibility that the cause may fail to operate. The point here is a very simple one. An equation may fail to be deterministic because it incorporates a random contribution to the effect beyond the influences contributed by the specified causes. But this does not make the causes themselves indeterministic.

[17] This is according to M. S. Morgan ('Correspondence Problems and the History of Econometrics', Sept. 1985, MS, University of York), who claims that most of the modern interpretations date from their first formal appearance, or even before.

These remarks are not intended to suggest that there is no way to use the error terms to introduce probabilistic causes. There is, but the results are cumbersome. Both the errors-in-variables and the errors-in-equations interpretation can provide a way of approaching the problem. For the errors-in-variables, just think of the influence that the cause x_i contributes when it operates to produce x_j—i.e. $a_{ji} x_i$—as the 'true' value of the influence; and as zero, which it contributes when it fails to operate, as the sum of the true value and some error that occurs on those occasions. The errors-in-equations interpretation will serve as well. Imagine, for example, someone who is going shopping. Normally their budget is a principal cause of their expenditure. But on occasion whim intervenes and the constraints of the budget are entirely offset. *Whim*, then, is an omitted factor that ought properly to be included in the equation; and its inclusion will make the budget behave like a purely probabilistic cause.

What is necessary to model probabilistic causes on either interpretation is that the effect of the error be equal in size to the influence of the given cause, and that it occur with fixed probability. This suggests that each term, $a_{ji} x_i$, in the exact, deterministic equation for x_j be replaced by a term of the form $a_{ji} x_i (1 - u_{ji}(x_i))$, where u_{ji} takes values 0 or 1. The resulting structure looks like this:

$$x_1 = u_1$$
$$x_2 = a_{21}x_1(1 - u_{21})$$
$$x_3 = a_{31}x_1(1 - u_{31}) + a_{32}x_2(1 - u_{32})$$
$$\cdots$$

With the assumption that the *u*s take only the values of 0 or 1, this kind of model can adequately represent the operation of purely probabilistic causes. But it should be noted that the resulting structure is no longer linear between the variables and the errors, since the errors—$a_{ji} x_i u_{ji}$—are functions of the exogenous variables in each equation. The need for this kind of interaction between the errors and the variables arises because I am trying to model a very special concept of probabilistic causality, a concept according to which the cause either contributes its entire influence or it does not contribute at all. A more general concept of a probabilistic cause may allow the size of the influence to vary in different ways from occasion to occasion. If it varies in a random way around its mean, the result is equivalent to the conventional structures, where the corresponding errors are independent of the level of the explanatory variable. I wish to treat the more restricted concept, not only because it is the one usually treated in the philosophical literature, but also because it picks out a kind of probabilistic causality that I think occurs familiarly in the world around us.

Although the job of representing probabilistic causes can be done by orchestrating the error terms, a different notation will provide a far simpler and more perspicuous picture of what is going on. The notation uses the simple device of including a factor that represents directly whether a cause operates or not. The new factors are designated by $\hat{a}$s (for action). For each cause x_i which contributes to the effect x_j, the idea is to introduce a new random variable $\hat{a}_{ji}$. The new variable is marked with a 'hat' because it shares many of the characteristics of indicator functions, which are conventionally represented in that way. Like an indicator function, $\hat{a}_{ji}$ takes on two values: it has value 1 if x_i operates to produce its customary influence, now designated with Greek coefficients α_{ji}, as in $\alpha_{ji}x_i$; and it takes value 0 when x_i fails to operate. This new variable is meant to represent a genuine physical occurrence, just the kind of occurrence that is presupposed in the concept of non-deterministic causes. But in those cases where the causes are purely probabilistic, it will coincide with no further physical state or property of the system that determines whether the cause operates or not. In most cases, though, there will be further empirical signs, beyond the mere occurrence of the effect itself, that will indicate whether the cause has operated or not.

Before considering what kinds of probabilistic relation these new variables have to each other and to the other more familiar variables, return for a moment to the conventional error terms. Using the $\hat{a}$ variables to represent the fundamentally probabilistic nature of the causes leaves the error terms free to play another role. The role is suggested by the case of radioactive decay. In treating decay, quantum physics employs two different concepts—that of stimulated emission and that of spontaneous emission. Stimulated emissions have some assignable external cause (but it is important to keep in mind that in all cases the causes in question are purely probabilistic); spontaneous emissions are random and uncaused events.[18] By analogy, the 'error terms' may be taken to represent

[18] This view obviously involves making a distinction between taking the decay product as the effect (its cause is the radioactive nucleus) and taking the emission of the decay product by the nucleus as the salient effect (an effect that has no cause at all).

purely spontaneous or uncaused occurrences of the effect in question, a kind of random background against which the proper causes operate. In this case the u_js (like the $\hat{a}_{ji}$s) represent genuine physical happenings—the spontaneous occurrence of x_j. But again, there need be no physical state or property which determines whether a u-type event happens or not, although (just as with the $\hat{a}$s) in many cases there will be independent ways to confirm that it has done so.

This interpretation has a considerable advantage over the 'ignorance' interpretation of the us, because it makes the usual independence assumptions about them perfectly natural. In general it is a fortunate accident—an accident that one has little reason to hope for—should the unknown causes be distributed so that they are probabilistically independent of the known causes. But if u stands for an event of spontaneous production, it is reasonable to suppose that this event occurs 'totally randomly'. The independence assumptions are one way to formulate this supposition.

The proposal, then, is to begin with equations that are one step back from the conventional ones:

Structure with probabilistic causes

$$x_1 = u_1$$
$$x_2 = \hat{a}_{21}\alpha_{21}x_1 + u_2$$
$$x_3 = \hat{a}_{31}\alpha_{31}x_1 + \hat{a}_{32}\alpha_{32}x_2 + u_2$$
$$\ldots$$

Recall that the multiplicative constants, which measure the size of the influence contributed, are represented here with αs; $\hat{a}$s appear with hats to represent the action of the influence. The equations are intended to have the same causal interpretation as before. Since $\hat{a}_{ji}$ has either the value 0 or the value 1, $\text{Prob}(\hat{a}_{ji}) = \text{Exp}(\hat{a}_{ji})$, and the assumption that the expectation is 0 means that x_i is not really a cause of x_j.

Besides introducing the possibility for causes to operate purely probabilistically, this scheme has another advantage: it separates the role of the multiplicative constants, which describe the strength of the influence's contribution, from the question of whether the influence is contributed at all. In the new scheme the parameters, α_{ji}, can always be assumed to be non-zero. Questions of causality depend on the $\hat{a}$s.[19] This distinction will be maintained from now on; questions about which causes genuinely appear in a equation for a given effect will henceforth be questions about which of the $\hat{a}_{ji}$ are universally (or 'almost always') zero. This will be abbreviated thus: $\hat{a}_{ji} = \perp$; if the operation is universally present when the cause is, the cause becomes deterministic. This will be expressed by $\hat{a}_{ji} = T$.

Before proceeding to a comparison between this new formalism and the more conventional one, I would like to turn to an unfinished problem, which can at last be treated now that the operation of causes has been introduced. Formula CC of Chapter 2 proposes to test a claim that C causes E by holding fixed a *complete set* of E's other causes. What makes a set *complete*? When all causes are deterministic, any set of factors which are separately necessary and jointly sufficient for the effect will be complete in the relevant sense. But in a case where causes are purely probabilistic, no disjunction of causes, however long, will be sufficient. The concept of operation

---

[19] The need for this kind of distinction is particularly clear when dealing with yes–no events of the kind Mackie studies. Consider for instance the derivation by D. Papineau in his 'Probabilities and Causes', *Journal of Philosophy*, 82 (1985), 57–74. Papineau takes over the Mackie formulation of section 1.3 here: $E \equiv AX \vee BY \vee CZ$. For Papineau, C will be a genuine partial cause just in case $Z \neq \perp$, or, more accurately for the purposes of his proof, just in case $C \nrightarrow \neg Z$, that is, if and only if C and Z can sometimes co-occur. But Z for him represents not the operation of the cause; instead it is supposed to represent some helping factors, factors that are necessary in order for the cause to operate. So his notation fails to distinguish the case in which C never brings about an E because it does not have the capacity to do so from cases where it indeed has the capacity but never has the opportunity because it is never present at the same time as its necessary helping factors. It is information about the former that we want to learn and to express, since in a good many cases we are in a position to rearrange the opportunities for the two factors to occur together.

The proof that Papineau gives establishes a connection between probabilities and inus conditions. The connection is the analogue for propositions or yes–no variables of the identifiability proofs described in section 1.2. For comparison with the conclusions there and also with formula CC, I give here a version of Papineau's proof. Given that $E \equiv AX \vee Y$ and given that the situation S is one in which A is probabilistically independent of Y (for instance because either $S \rightarrow Y$ or $S \rightarrow \neg Y$, i.e. Y is held fixed in S), then (i) if $P(E/A) > P(E/\neg A)$ it follows that $A \rightarrow X \neq \perp$; and (ii) if $A \rightarrow X \neq \perp$ and $A \rightarrow (X \nrightarrow Y)$ it follows that $P(E/A) > P(E/\neg A)$. The proof follows almost immediately upon expansion: $P(E/A) = P(AX \vee Y/A) = P(X/A) + P(Y/A) - P(XY/A)$. Given the independence of Y and A, this is greater than $P(E/\neg A) = P(Y/\neg A)$ iff $P(X/A) > P(XY/A)$. Papineau's own arguments for giving a causal interpretation to the inus conditions thus identified are quite different from either my use of the open-back-path condition or my proposal to begin only with causal factors. They can be found in his 'Causal Asymmetry', *British Journal for the Philosophy of Science*, 36 (1985), 273–89.

enters naturally here. A complete set of causes for an effect E is a set of causes of E such that (i) if E occurs some member of that set is bound to have occurred and to have operated, and (ii) if any member of that set occurs and operates, and no preventatives of E (that is, in this simple framework, causes of $\neg E$) occur at the same time or during the relevant period after, then E occurs. Using the concepts both of a preventative and also of the operation of a cause, the notion of a complete set is easy to formulate. Otherwise I do not know how to characterize it. Since the idea of a complete set of causes is necessary to arguments like those of Chapter 1, which aim to justify our usual probabilistic measures for causality, I take this to provide one more strong argument for the place of non-Humean concepts in our familiar scientific picture of the world.

3.3.2 The New Formalism and the Old

Turn now to the important question of how the $\hat{a}$s relate to the other variables. They do, after all, represent physical happenings. What associations do the events they represent have with others? That is, how does the operation of a given cause bear on the operation of others, or on other different kinds of event?[20] The answer to this question brings out an important fact about how the new formalism is connected with the old. The reason for introducing the $\hat{a}$ variables is to make explicit the indeterministic nature of the causes. But sometimes this information is irrelevant to the question of central concern here—how to use probabilities to pick out causes. In particular, this is the case if all causes operate completely 'randomly'; that is, if all the operations are probabilistically independent of everything except their own consequences and of any descendants of these. When the operations satisfy this strong independence requirement, the new formalism and the old will determine the causes in exactly the same way: the expanded system of equations, which includes the action variables $\hat{a}_{ji}$, will generate all the same probabilistic criteria as the conventional fixed-parameter equations. The parameters of the conventional equations simply combine the new multiplicative

---

[20] One assumption has already been built into the notation. In principle, the probability that a cause operates may well depend on how much or how little of the cause is present: the probability that x_i operates to produce x_j may depend on the level of x_i. For simplicity, I consider here only cases in which the probability of producing an influence is independent of the level of the cause. This assumption does not of course affect the *amount* of influence, which continues—through the αs—to be proportional to the level of the cause.

constants α_{ij} with the expectations of the $\hat{a}_{ji}$. It works this way because the $\hat{a}_{ji}$ factor out in any probabilistic calculation when they are independent of everything else, and their probabilities appear as multiplicative constants, just as if they were fixed parameters. This is why the new equations were described in the last part as 'one step back' from the conventional ones.

As an illustration, consider again the derivation of the common-cause condition of Chapter 1. Given total independence of the $\hat{a}$s, the condition looks exactly the same in the new scheme as in the original, so long as the ordinary assumptions about the independence and scaling of the error terms are made. Given a three-variable model:[21]

$$x_1 = u_1$$
$$x_2 = \hat{a}\alpha x_1 + u_2$$
$$x_3 = \hat{b}\beta x_1 + \hat{c}\gamma x_2 + u_3$$

calculate, as before,

$$\begin{aligned} \text{Exp}\,(x_2 x_3/x_1) &= \beta x_1\,\text{Exp}\,(\hat{b}x_2/x_1) + \gamma\,\text{Exp}\,(\hat{c}x_2^2/x_1) \\ &= bx_1\,\text{Exp}\,(x_2/x_1) + c\text{Exp}\,(x_2^2/x_1) \end{aligned}$$

and

$$\begin{aligned} \text{Exp}\,(x_3/x_1) &= \beta x_1\,\text{Exp}\,(\hat{b}/x_1) + \gamma\,\text{Exp}\,(\hat{c}x_2/x_1) \\ &= bx_1 + c\,\text{Exp}\,(x_2/x_1) \end{aligned}$$

where the abbreviations $b = \beta\text{Exp}\,\hat{b}$ and $c = \gamma\text{Exp}\,\hat{c}$ have been used. Given that $u_2 \neq 0$ so that the factor in the denominator is not zero, and assuming that all multiplicative parameters are also non-zero, the result is familiar:

Common-cause condition 1

$\hat{c} = \perp$ iff $\text{Exp}\,(x_2 x_3/x_1) = \text{Exp}\,(x_2/x_1)\,\text{Exp}\,(x_3/x_1)$.

Thus the complete independence of the action variables reduces the new system of equations to the standard ones, and thereby generates the familiar criteria for causality.

But independence is not always an appropriate assumption to make. Correlatively, the familiar criteria are not always the right ones to use. A typical case occurs when a cause operates subject to constraints, so that its operation to produce one effect is not

---

[21] The derivation assumes that the expectations of $\hat{a}$ and $\hat{b}$ are independent of the level of x_1 and similarly the expectation of $\hat{c}$ is independent of x_2.

independent of its operation to produce another. For example, an individual has \$10 to spend on groceries, to be divided between meat and vegetables. The amount that he spends on meat may be a purely probabilistic consequence of his state on entering the supermarket; so too may be the amount spent on vegetables. But the two effects are not produced independently. The cause operates to produce an expenditure of n dollars on meat if and only if it operates to produce an expenditure of $10 - n$ dollars on vegetables. Other constraints may impose different degrees of correlation.

The first probabilistic causes were represented with urn models by Jakob Bernoulli, and these models can still be of help in understanding causes and their constraints. The simple case where a cause operates independently to produce its different effects can be modelled on independent drawings from separate urns. The urns contain black and white balls in the appropriate ratios. The cause operates to produce the first effect if and only if a black ball is drawn from the first urn; and the second, if and only if a black ball is drawn from the second. In the more general case, there is a single urn containing balls yoked together in pairs, where the four different kinds of pair appear in the appropriate ratios. A single drawing is made to determine simultaneously the production of the first effect and the production of the second.

When correlations like these are admitted, the conventional probabilistic criteria by which causes are determined will be changed. To illustrate, consider the question of factorizability again, this time allowing that x_1 may be constrained in its production of the two effects, x_2 and x_3, i.e. allowing for correlation between $\hat{a}$ and $\hat{b}$. Otherwise keep all the same assumptions as before. In this case Exp $(\hat{b}x_2/x_1)$ must be recalculated; for $\hat{b}$ is no longer independent of $x_2 = \hat{a}x_1 + u_2$. By hypothesis Exp $(\hat{a}\hat{b}) \neq$ Exp $(\hat{a})$ Exp $(\hat{b})$, so factorizability fails, and the appropriate common cause condition becomes:

Common-cause condition 2

$$c = \perp \text{ iff Exp } (x_2x_3/x_1) = \text{ Exp } (\hat{a}\hat{b})\alpha\beta x_1{}^2$$

This is the natural condition. It just says that, when x_2 contributes nothing to x_3, their joint expectation is the expectation of x_1's joint contribution to each.[22]

---

[22] The only time that this common-cause condition fails is when the denominator of the preceding equation becomes zero, and that will not happen so long as $u_2 \neq 0$. Even when $u_2 = 0$ the denominator becomes zero only when $\hat{b}$'s occurrence guarantees $\hat{a}$'s (i.e. $\hat{b} \rightarrow \hat{a}$). Otherwise the common-cause condition is valid.

Similar calculations provide criteria for $\hat{a}$ and $\hat{b}$ as well, and the method is easily generalized to more than three variables. The general lesson is that, even in the extended scheme where causes may be probabilistic, the causes may still be determined from the probabilities. But the probabilities are no longer just those for the occurrence of various measurable quantities; the operations must be treated as well. It is not enough to know how often the cause and effect co-occur, how often the joint effects co-occur, and the like. One must know how often the causes operate as well, and how often the operations coincide. Still, these are probabilities for genuine physical events, even though they are not ones favoured by the purest of Humeans.

3.3.3. *A Lesson for Quantum Mechanics*

Before closing this section, I want to return to the original notation in which probabilistic causes are represented as errors-in-variables, in order to make a point that bears on recent questions about hidden variables in quantum mechanics. The questions focus on an experiment originally proposed in 1935 by Einstein and two collaborators, B. Podolsky and N. Rosen.[23] The issues surrounding the Einstein–Podolsky–Rosen (EPR) experiment will be discussed in detail in Chapter 6; but it is already possible to summarize the main point here. When the operation of causes is represented explicitly, correlations between the operations can be introduced directly, as probabilistic constraints on the operations. In the errors-in-variables notation these will appear as correlations between the errors. Econometrics frequently uses another method: correlations among the errors are introduced by imposing constraints on the explanatory variables, often in the form of some linear relation among them. Equilibrium equations are one familiar kind of example; everyone knows something about the classical models which set quantity demanded equal to quantity supplied.

Consider, for example, the most simple case of a common-cause model with a linear constraint between the two effect variables. It will not be necessary to go beyond that to see an important point about the EPR experiment.

---

[23] A. Einstein, B. Podolsky, and N. Rosen, 'Can a Quantum Mechanical Description of Physical Reality be Considered Complete?', *Physical Review*, 47 (1935), 770–80.

Common-cause model with constraint (I)

$$x_1 : \text{exogenous}$$
$$x_2 = \alpha x_1 (1 - u)$$
$$x_3 = \beta x_1 (1 - v)$$
$$x_3 = \delta x_2$$

The last equation expresses the constraints in structural form; often weaker constraints, just involving probabilities, are used instead (for instance, $\text{Exp}(x_3 x_2) / \text{Exp}(x_2^2) = \delta$) . The system is consistent only if $v = 1 - \alpha\delta/\beta + (\alpha\delta/\beta)u$, in which case v and u can be represented using a common factor, $\hat{c}$: $v = \hat{c} + \hat{n} - \hat{c}\hat{n}$ and $u = \hat{c} + \hat{m} - \hat{c}\hat{m}$, where $\hat{m} = (\alpha\delta - \beta + \beta\hat{n})/\alpha\delta$. We may say that u and v *share* a common factor since (recalling that u and v take only the values 0 and 1) in the notation of Boolean logic the expressions above for u and v are equivalent to $v \equiv \hat{c} \vee \hat{n}$ and $u \equiv \hat{c} \vee \hat{m}$.

To make the connection with the notation in which the operations are represented directly, recall that a given cause operates if and only if its associated error term takes on the value zero. So, letting $\hat{\phi} \equiv \neg \hat{c}$ (or $\hat{\phi} = 1 - c$), $\hat{s} \equiv \neg \hat{m}$, and $\hat{t} \equiv \neg \hat{n}$, the equations of structure I can be recast as a *local model*:

II
1. x_1 : exogenous
2. $x_2 = \alpha(\hat{\phi} \cdot \hat{s})x_1$
3. $x_3 = \alpha(\hat{\phi} \cdot \hat{t})x_1$

To see why models of form II warrant the description 'local', compare structures of form I with structures of form II. By hypothesis x_2 precedes x_3—that is assumed throughout. In general the events represented by these variables will be separated not only in time but in space as well. In neither model does x_2 cause x_3. Yet values of the two are related. How do they come to be related? Structure I simply asserts the brute fact that the two variables are functionally related to each other. Structure II supposes that this functional relation is not just a brute fact, but has a simple causal account. It arises because the operation of x_1 to produce x_2 overlaps its operation to produce x_3: the one event always shares some part with the other.[24] The relationship between the separated occurrences x_2 and x_3 is due to facts about how two events that occur together at just the same

---

[24] For an account of events and their parts, see J. J. Thompson, *Acts and Other Events* (Ithaca, NY: Cornell University Press, 1977).

time and place—viz. x_1's operation to produce x_2, and x_1's operation to produce x_3—relate to each other. That is the sense in which the model is local.

As a simple illustration, consider again the case of the shopper with a fixed budget: $10 to spend on both meat and vegetables, where the shopper's state of mind on entering the supermarket is supposed to be a probabilistic cause of the amounts spent. The meat is picked up first, the vegetables several minutes later. Yet the amounts are correlated. One may view the correlation along the lines of structure I as a brute fact: the separated events just are related to one another. Or, they can be modelled locally, by assuming that the decision to buy y dollars worth of meat is the very same event as the event of deciding to buy $(10 - y)$ dollars worth of vegetables. In that case the model will look like structure II.

Does every probabilistic structure with consistent linear constraints have a local equivalent in the same variables? In general, no; but in some special cases, the answer is yes. In particular, if the constraint involves only variables which have all their causes in common, the model can always be recast as a local one. Structure I is a particularly simple example of this. So too is the structure of the Einstein–Podolsky–Rosen experiment.

The EPR experiment is concerned with correlations between the outcomes of measurements on separated systems, originally produced together at a common source. In modern-day versions, the measurements determine the spin along specific directions of two particles prepared at the source to be in a special state, called the singlet state. Let $x_l(\theta)$ represent the outcome for a measurement of spin in the direction θ in the left wing of the experiment; $x_r(\theta')$, the outcome for the (possibly different) direction θ' in the right wing; and designate the action function for the occurrence of the quantum singlet state by $\hat{\Phi}s$. Both $x_l(\theta)$ and $x_r(\theta')$ may take either the values 1 (for spin up in the direction θ or θ') or 0 (for spin down). Phenomenologically the experimental situation looks like this:

EPR
$$x_l(\Theta) = \hat{\Phi}s(1 - u(\Theta))$$
$$x_r(\Theta') = \hat{\Phi}s(1 - v(\Theta'))$$
$$\text{Exp}(x_l(\Theta)x_r(\Theta')) = \tfrac{1}{2}\sin^2\{(\Theta - \Theta')/2\}$$
$$\text{Exp}(u(\Theta)) = \tfrac{1}{2}$$
$$\text{Exp}(v(\Theta')) = \tfrac{1}{2}$$

I think the real lessons of the experiment concern quantum realism. But sometimes the results are taken to bear on causality. The question then is, can the correlations between the outcomes in this experiment be derived from a local common-cause model? Evidently, yes. As written here, the EPR structure is already a common-cause model, with consistent constraints, and any such model is trivially equivalent to a local one. This means that there is nothing in the probabilities to show that EPR cannot have a common-cause structure, and also a structure in which there are no correlations between the actions of separated causes.

Perhaps one wants more from a causal structure than just getting the probabilities right. It is usual, for instance, to demand some kind of spatio-temporal contiguity between the cause and the effect. Whether that is possible in the case of quantum mechanics will be discussed in Chapter 6. But with respect to the probabilities alone, there is no problem in assuming a common cause for the separated measurement outcomes. This structure only looks to be impossible if one uses the wrong criterion for a common cause; i.e. if one fallaciously uses condition 1, which is appropriate to models without constraints, rather than condition 2, which is the right one for the EPR experiment. Chapter 6 asks, 'What can econometrics teach quantum physics?' The answer lies in the straightforward reminder that equilibrium conditions put implicit restrictions on the error processes; and when the error processes are not independent, the causal structure cannot be determined in the ordinary way.

3.4. More in Favour of Singular Causes

With this apparatus in place, the intuitive arguments of section 3.2 can be recast more formally. I do so as a double-check. Each method of argument has its own internal weaknesses; together the two serve to balance each other. The reader who is not interested in the formulae can scan quickly for the principal philosophical claims.

Because the examples in section 3.2 follow recent philosophical literature in discussing qualitative causal relations, rather than quantitative ones, a Boolean representation will be more appropriate than one using linear equations. Mackie's treatment is the guide, except that his inus account requires that each complete cause be sufficient for its effect. Following the methods of the last section,

Mackie's deterministic causes can be turned into probabilistic ones by introducing a proposition that indicates whether the cause operates or not. As in the last section, these propositions will be designated by 'hats': $\hat{a}$, $\hat{b}$,. . . .

A simple example like that of the birth-control pills, which involves one cause with dual capacities operating against a fixed background, will have the structure given in Model M.

Model M

$$C' = \hat{a}C \vee \hat{b}B$$
$$\neg P = \hat{c}C \vee \hat{d}B$$
$$E = \hat{e}C' \vee \hat{f}P$$

Here C is the dual cause (contraceptives) which can either promote E (thrombosis) by producing a later cause C' (chemical in blood) or inhibit E by preventing a cause that might occur later, designated by P (pregnancy). B (for background) summarizes the effects of all other factors simultaneous with C that can also produce P or C'. In addition, it is assumed that nothing else is relevant.

The strategy that Eells and Sober take to eliminate singular causes is to control for B. They then judge the causal role of C by comparing the probability of E with and without C, when B is fixed. The two probabilities in this case are given by formulae X_1 and X_2:

X

1. Given B, $P(E/C) = P(\hat{e}\hat{a} \vee \hat{e}\hat{b} \vee \hat{f}\neg\hat{d}\neg\hat{c}/C)$
2. Given B, $P(E/\neg C) = P(\hat{e}\hat{b} \vee \hat{f}\neg\hat{d}/\neg C)$

It is apparent from these formulae that the relation between $P(E/C)$ and $P(E/\neg C)$ will depend on exactly what values the probabilities take. There is nothing in the structure of the formulae that decides the matter. This result duplicates the conclusion of section 3.2. If only B is held fixed, anything can happen: the probability of the effect may go either up or down in the presence of the dual cause; it may even stay the same.

In section 3.2 this was accounted for in a simple way: holding fixed B produces a kind of averaging. It averages over populations in which some alternative causes—here P and C'—would naturally occur and ones in which they would not. In Model M it is $\hat{b}$ that tells whether C' would naturally occur independently of C; similarly, $\hat{d}$ tells about the independent occurrence of P. The averaging is apparent in formulae X_1 and X_2, which condition on neither $\hat{d}$ nor $\hat{b}$. As a consequence the populations picked out for examination are mixed ones, where $\hat{b}$ and $\hat{d}$ sometimes occur and sometimes do not, rather than the more homogeneous subgroups in which $\hat{b}$ and $\hat{d}$ are fixed. It is evident that this will make a difference since, as the formulae show, both $\hat{b}$ and $\hat{d}$ matter to E.

My proposal here is to look instead at the frequency of E by considering *four separate populations* in turn, just as one would do if B produced its effects on C' and P a little before C acted. The populations, then, are segregated according to the values of $\hat{b}$ and $\hat{d}$: two variables, each of which may be either T or $\perp$, yield four populations. In each population the strategy is to assume that B is given and then use formulae X_1 and X_2 to examine the influence of C on E.

The *first* population is chosen so that every individual has C' and would have P from the action of B alone unless C acts to prevent it. That is, a population of those individuals for which $\hat{b} = T$ and $\hat{d} = \perp$. In this population C could only prevent E; whether it has the power to do so or not depends on $\hat{c}$ and $\hat{f}$: it can do so if and only if $\hat{c}$ and $\hat{f}$ both occur sometimes, i.e. $\hat{c} \neq \perp$ and $\hat{f} \neq \perp$.[25] But these are exactly the same conditions that guarantee that $P(E/\neg C) > P(E/C)$, since in the case where $\hat{b}$ occurs and $\hat{d}$ does not, formula X_1 and X_2 reduce to $P(E/\neg C) = P(\hat{e} \vee \hat{f})$ and $P(E/C) = P(\hat{e} \vee \hat{f}\neg\hat{c})$. So the probability of $\hat{e} \vee \hat{f}$ is bound to be bigger than that of $\hat{e} \vee \hat{f}\neg\hat{c}$ so long as $\hat{f} \neq \perp$ and $\neg\hat{c} \neq T$.[26] Here matters are arranged as they should be; the probability for E will decrease with C if and only if occurrences of C do prevent Es in this situation.

In the *second* population, where B already acts both to produce C' and to prevent P, C will be causally irrelevant, and that too will be shown in the probabilities. For $P(E/C) = P(\hat{e}\hat{a} \vee \hat{e}) = P(\hat{e}) = P(E/\neg C)$, when $\hat{b} = T$ and $\hat{d} = T$.

In the *third* population, C can both cause and prevent E, and the probabilities can go either way. This group provides no test of the powers of C. The *fourth* population, where B would by itself already prevent P but not cause C', is just the opposite of the first. The

[25] More precisely, if and only if $\hat{b} = T$ and $\hat{d} = \perp$ and $C = T \rightarrow \hat{f} \neq \perp$ and $\hat{c} \neq \perp$.

[26] And $\hat{f} \nrightarrow \neg c$; but this is precluded by the locality assumption. Since $\hat{f}$ and $\hat{c}$ represent actions of separated causes. In fact, it is required that $\hat{b} = T$ and $\hat{d} = \perp$ and $C = T \nrightarrow \hat{f} \neg \hat{c}$, and also that $\hat{b} = T$ and $\hat{d} = \perp$ and $C = \perp \nrightarrow \hat{f} = \perp$. But these are equivalent to the condition stated given the independence assumptions for local models. It is assumed throughout that all models are local.

only possible influence C could have is to cause E, which it will do if $\hat{e}\hat{a} \neq \perp$; and, as wanted, $\hat{e}\hat{a} \neq \perp$ if and only if $P(E/C) > P(E/\neg C)$.

What does this imply about the possibility of testing causal claims? On the one hand the news is good. Although the analysis here is brief, and highly idealized, it should serve to show one way in which causal claims can be inferred from information about probabilities. A cause may have a number of different powers, which operate in different ways. In the simple example of the contraceptives, the cause has two routes to the effect—one positive and one negative. But this case can be generalized for cases of multiple capacities. The point is that, for each capacity the cause may have, there is a population in which this capacity will be revealed through the probabilities. But to pick out that population requires information not only about which of the other relevant factors were present and which were absent. It is also necessary to determine whether they acted or not. This is the discouraging part. It is discouraging for any investigator to realize that more fine-grained information is required. But the conclusion is a catastrophe for the Humean, who cannot even recognize the requisite distinction between the occurrence of the cause and its exercise.

Recall that the discussion in section 3.2 of how one might try to circumvent the need for this distinction and for singular causes was interrupted. It is time to return to it now, and to take up the two remaining proposals, one involving techniques of path analysis, the other the chopping of time into discrete chunks.

3.4.1. Strategy (iii)

Path analysis begins with graphs of causal structures, like Fig. 3.3 for the contraceptive example. Claims about any one path are to be tested by populations generated from information about all the other paths. The proposal is to hold fixed some factor from each of the other paths, then to look in that population to see how C affects the probability of E. If the probability increases, that shows that the remaining path exists and represents a positive influence of C on E; the influence is negative if the probability decreases; and it does not exist at all when the probability remains the same.

The idea is easy to see in the idealized contraceptive example, where only two paths are involved. In the population of women who have definitely become pregnant by time t_2—or who have definitely failed to become pregnant—the power of the contraceptives to prevent thrombosis by preventing pregnancy at t_2 is no longer relevant. Thus the positive capacity, if it exists, will not be counterbalanced in its effects, and so it can be expected to exhibit itself in a net increase in the frequency of thrombosis. Similarly, holding fixed C' at t_2 should provide a population in which only the preventative power of the contraceptives is relevant, and hence could be expected to reveal itself in a drop in the number of cases of thrombosis. This is easy to see by looking at the formulae

> Y
>
> 1. Given B, $P(E/CP) = P(\hat{e}\hat{a} \vee \hat{e}\hat{b} \vee \hat{f}/CP)$
> 2. Given B, $P(E/\neg CP) = P(\hat{e}\hat{b} \vee \hat{f}/\neg CP)$
>
> and
>
> Z
>
> 1. Given B, $P(E/C\neg P) = P(\hat{e}\hat{a} \vee \hat{e}\hat{b}/C\neg P)$
> 2. Given B, $P(E/\neg C\neg P) = P(\hat{e}\hat{b}/\neg C\neg P)$

When P is held fixed, $\hat{f}\hat{c}$, which represents the track from C to E via P, does not enter the formula at all. So long as all operations are independent of all others, it follows, both in the case of P and in the case of $\neg P$, that $P(E/C) > P(E/\neg C)$ if and only if $\hat{e}\hat{a} \neq \perp$; that is, if and only if there really is a path from C to E, through C'.

When the operations are not independent, however, the situation is very different; and the conventional tactics of path analysis will not work. The practical importance of this should be evident from the last section. Some independence relations among the operations will be guaranteed by locality; and it may well be reasonable, in these typical macroscopic cases, to insist on only local models. But it is clearly unreasonable to insist that a single cause operating entirely locally should produce each of its effects independently of each of the others. Yet that is just what is necessary to make the procedures of the third strategy valid.

Consider what can happen in Model M when there is a correlation between the operation of C to prevent pregnancy and its operation to produce the harmful chemical. In particular, assume that if C fails in a given individual to prevent pregnancy, it will also fail to produce the chemical (i.e. $\neg \hat{c} \rightarrow \neg \hat{a}$). Behaviour like this is common in decay problems. For instance, consider the surprising case of protactinium, finally worked out by Lise Meitner and Otto Hahn. The case is surprising because the same mother element, protactinium, can by two different decay processes produce different final elements—in one case uranium, plus an electron, plus a photon; in the other thorium, plus a positron, plus a photon. The two processes are entirely distinct—the first has a half-life of 6.7 hours, the second of 1.12 minutes; and, as is typical in decay processes, the effects from each process are produced in tandem: the protactinium produces the uranium if and only if it produces an electron as well; similarly, thorium results if and only if a positron is produced.

Imagine, then, that the contraceptives produce their effects just as the decaying protactinium does. What happens to the probabilities? Consider first the $+P$ population. Expanding Y_1 and Y_2 gives

Y_1' : Given B, $P(E/CP) = P(\hat{e}\hat{a} \vee \hat{e}\hat{b} \vee \hat{f}/\neg\hat{d}\neg\hat{c}C)$
Y_2' : Given B, $P(E/\neg CP) = P(\hat{e}\hat{b} \vee \hat{f}/\neg\hat{d}\neg C)$

Given the other independence assumptions appropriate to a local model, it is apparent that $P(E/CP) = P(E/\neg CP)$ when $\neg\hat{c} \rightarrow \neg\hat{a}$, and this despite the fact that $\hat{e}\hat{a} \neq \perp$. In this case a probability difference may still show up in the $\neg P$ population; but that too can easily disappear if further correlations of a suitable kind occur.[27] The point here is much the same as the one stressed in the last section. Many of the typical statistical relations used for identifying causes—like the path-analysis measure discussed here or the factorizability condition discussed in the last section—are appropriate only in local models, and then only when local operations are independent of each other. When causes with dual capacities produce their effects in tandem, steps must be taken to control for the exercise of the other capacities in order to test for the effects of any one. If you do not control for the operations, you do not get the right answers.

Besides these specific difficulties which arise when operations correlate to confound the probabilistic picture, there are other intrinsic problems that block the attempt to use causal paths as stand-ins for information about whether a power has been exercised

[27] When $\neg\hat{c} \rightarrow \neg\hat{a}$, given the other assumptions of a local model, it will follow that $P(\hat{e}\hat{a} \vee \hat{e}\hat{b}/C(\hat{c} \vee \hat{d})B) = P(E/C\neg PB) = P(E\neg C\neg PB) = P(\hat{e}\hat{b}/\neg C\hat{d}B)$ so long as $(P(\hat{c})/P(\hat{d}) - P(\hat{c}))x = y+z$, where $x = P(\hat{b}\hat{d})$, $y = P(\hat{b}\neg\hat{d})P(\hat{c})$, $z = P(\hat{a}\neg\hat{b})P(\hat{c})$.

or not. A detailed consideration of these problems is taken up in joint work by John Dupré and me.[28] I here summarize the basic conclusions. The first thing to note about adopting causal paths as a general strategy is that there is some difficulty in formulating precisely what the strategy is. I have here really given only an example and not a general prescription for how to choose the correct test populations. It turns out that a general formulation is hard to achieve. Roughly, to look for a positive capacity one must hold fixed some factor from every possible negative path, and these factors must in addition be ones that do not appear in the positive paths. This means that to look for a positive capacity one has to know exactly how each negative capacity can be exercised, and also how the positive capacity, if it did exist, would be exercised; and vice versa for finding negative capacities.

This situation, evidently epistemologically grotesque, becomes metaphysically shaky as well when questions of complete versus partial causes are brought forward. For it then becomes unclear what are to count as alternative causal paths. It may seem that the question has a univocal answer so long as one considers only complete causes. But a formulation that includes only complete causes will not have very wide application. It is far more usual that the initial cause be only a partial cause. Together with its appropriate background, it in turn produces an effect which is itself only a partial cause of the succeeding effect relative to a new background of helping factors. To delineate a unique path, the background must be continually revised or refined as one passes down the nodes, and what counts as a path relative to one revision or refinement will not be a possible path relative to another.

Perhaps these problems are surmountable and an acceptable formulation is possible; but I am not very optimistic about the project on more general grounds. We are here considering cases where a single causal factor has associated with it different opposing capacities for the same effect. What are we trying to achieve in these cases by holding fixed intermediate factors on other causal paths? The general strategy is to isolate the statistical effects of a single capacity by finding populations in which, although the cause may be present, its alternative capacities are capable of no further exercise. Then any difference in frequency of effect must be due to the hypo-

[28] J. Dupré and N. Cartwright, 'Probability and Causality: Why Hume and Indeterminism Don't Mix', *Nous*, 22 (1988), 521–36.

thesized residual capacity. The network of causal paths is a device that is introduced to provide a way to do this within the confines of the Humean programme. To serve this purpose the paths must satisfy two separate needs. On the one hand, they must be definable entirely in terms of causal laws, which can themselves be boot-strapped into existence from pure statistics. On the other hand, the paths are supposed to represent the canonical routes by which the capacities operate, so that one can tell just by looking at the empirical properties along the path whether the capacity has been exercised. I am sceptical that both these jobs can be done at once.

It should be noted that this scepticism about the use of causal paths is not meant to deny that capacities usually exercise themselves in certain regular ways. Nor do I claim that it is impossible to find out by empirical means whether a capacity has been exercised or not. On the contrary, I support the empiricist's insistence that hypotheses about nature should not be admitted until they have been tested, and confirmed. But the structure of the tests themselves will be highly dependent on the nature and functioning of the capacity hypo-thesized, and understanding why they are tests at all may depend on an irrevocably intertwined use of statistical and capacity concepts. This is very different from the Humean programme, a programme that distrusts the entire conceptual framework surrounding ·capa-cities and wants to find a general way to render talk about capacities and their exercise as an efficient, but imperspicuous, summary of facts about empirically distinguishable properties and their statis-tical associations. The arguments here are intended to show that the conventional methods of path analysis give no support to that programme.

Cases from quantum mechanics present another kind of problem for the methods of path analysis, problems which are less universal but still need to be mentioned, especially since they may help to illus-trate the previous remarks. The path strategy works by pin-pointing some feature that occurs between the cause and the effect on each occasion when the cause operates. But can such features always be found? Common wisdom about quantum mechanics says the answer is no. Consider the Bohr model of the atom, say an atom in the first excited state. The atom has the capacity to de-excite, and thereby produce a photon. But it will not follow any path between the two states in so doing. At one instant it is in the excited state, at the next, it is de-excited; and there are no features in between to mark the passage.

The absence of path is easily turned into a problem for those who want to render causes entirely in terms of probabilities. Imagine, for instance, that the spacing between the ground state and the first excited state is just equal to that between the first excited state and the second. This will provide an extremely simple example of an atom with dual capacities: in de-exciting, the atom can produce photons of the frequency corresponding to the energy spacing; but it can also annihilate them, simultaneously with moving to the higher state. In order to assure that either of these capacities will reveal itself in a change in the total number of photons, the operation of the other capacity must be held fixed. This does not mean that it must literally be controlled for, but some means must be discovered to find out about, and compute, the effects of overlaps between the two operations. A variety of experimental procedures may work; or, more simply, the results may be calculated from the quantum description of the atom and its background. What ties all these procedures together, and justifies them as appropriate, is the fact that they are procedures which can be relied on to find out what would happen if the exercise of one or the other of the two opposed capacities was held fixed.

Return now to the discussion of causal paths. The general strategy of path analysis is to substitute information about whether features on the path between cause and effect have occurred in an individual case for information about whether the cause has succeeded in making its contribution to the effect in that case. But in the Bohr atoms there are no paths. It becomes clear that it is the singular causal fact itself that needs to be held fixed, and looking at features along a path is one possible way of doing that; but in other cases it might not work. The difference is readily expressed by two different formulations. Path analysis always requires a model like M that interposes something between the putative cause and its effects. The point is easier to see in the case of linear equations than in Boolean relations, so I switch back to the notation of section 3.3. Consider formula N.

$$N$$

$$x_\mathrm{e} = \hat{g}\mu x_\mathrm{c} - \hbar\eta x_\mathrm{c} + w$$

Taking x_e as the effect in question, and x_c as the cause, what one really wants to know is whether $\hat{g}$, which represents a positive capacity of C to produce E, is impossible or not, i.e., does $\hat{g} = \perp$; or, alternatively, is the negative capacity genuine, i.e., does $\hat{h} = \perp$?

Path analysis reformulates the question by interposing the intermediate effects $x_{c'}$ and x_p:

$$O$$
$$x_{c'} = \hat{a}\alpha x_c + \hat{b}\beta x_b + u$$
$$x_p = -\hat{c}\gamma x_c + \hat{d}\delta x_b + v$$
$$x_e = \hat{e}\xi x_{c'} + \hat{f}\rho x_p + w$$

In this case the exercise of the positive capacity becomes a conjunction of the two intermediate operations: $\hat{g} = \hat{a}\hat{e}$; similarly for the negative capacity: $\hat{h} = \hat{c}\hat{f}$. The hope is to avoid the need to control for the operations by controlling instead for their intermediate consequences, x_p and $x_{c'}$. The earlier example showed how correlations and constraints can frustrate this hope. But the point here is different. It should be apparent that, even failing confounding correlations, the strategy will work only in domains where trajectories between the cause and effect are assured. More than that, it must be supposed that for each capacity there is some fixed and specifiable set of routes by which it exercises itself; and that is what I want to call into question. It seems to me that the plausibility of that suggestion depends on a thoroughgoing and antecedent commitment to the associationist view, that it must be the paths that are laid down by nature with capacities to serve as summarizing devices. But there is the converse picture in which the capacities are primary, and in which whatever paths occur, from occasion to occasion, are a consequence of the manner in which the capacities exercise themselves, in the quite different circumstances that arise from one occasion to another. Within this picture the paths are likely to be quite unstable and heterogeneous, taking on the regularity and system required by path analysis only in the highly ordered and felicitously arranged conditions of a laboratory experiment.

The contrast between N and O also brings out nicely the increased representational strength that comes with the notation of section 3.3. In the conventional fixed-parameter notation, N would become N'.

$$N'$$

$$x_e = \mu x_c - \eta x_c + w = \psi x_c + w$$

where $\psi = \mu - \eta$. In this case neither the positive nor the negative capacity would be truly represented, but just the net result. Prima facie, dual capacities disappear when fixed parameters substitute for the operational values. Causal intermediaries serve as a way to reintroduce them, as in Model O', which is the analogue of O:

$$O'$$

$$x_{c'} = \alpha x_c + \beta x_b + u$$
$$x_p = -\gamma x_c + \delta x_p + v$$
$$x_e = \xi x_{c'} + \phi x_p + w$$

Although the true causal role of *C* will be concealed in the reduced form, N', and even totally obliterated in the case where $\mu = \eta$, it reappears, and its dual nature becomes manifest, in the finer structure of O'. But the device depends essentially both on the existence of the causal intermediaries and on the felicity of the choice of time-graining. The grain must be fine enough to ensure that no cause represented at any one node of the path itself has opposed capacities with respect to its effect, represented at the next node. If it does, the fixed-parameter scheme will necessarily represent only its net result, and not its true causal nature; and that can give rise to the familiar difficulties. In the most flagrant case, the two opposing capacities may balance each other, and the path between the original cause and its effect will appear to be broken at that node, since the net result of that cause on its effect will be zero.

This is not meant to suggest that correct time-graining is impossible. On the contrary, I think that various domains of physics, engineering, biology, and medicine repeatedly succeed in building adequate models, and using something like a fixed-parameter representation, with no explicit reference to singular causings or the operations of capacities. Rather, the point is another version of the anti-Hume argument. What makes the model appropriate for representing causal structure is not just the truth of the equations but, in this case, the aptness of the time-graining as well. And what is and is not an appropriate grain to choose cannot be determined without referring to capacities and how they operate.

3.4.2. Strategy (iv)

These remarks bear immediately on the last strategy which I want to discuss for circumventing the need for singular causes, that is, the strategy to chop time into discrete chunks. I will discuss it only briefly, for I think the general problems it meets will now be easy to recognize. The idea of this strategy is to make the time-chunks so small that no external factors can occur between the cause and its immediate successor on the path to the eventual effect. The hope is to use a formula like CC from Chapter 2 to define a concept of *direct cause*, where C and E in the formula are taken to be immediate successors and the external factors that define the test population are all simultaneous with C. The more general concept of *cause simpliciter* is to be defined by an ancestral relation on direct cause.

The difficulty with this proposal is the evident one: in fact, time is not discrete, at least as far as we know. This does not mean that a discrete model is of no use for practical investigations, where a rich variety of concepts is employed. But it does make difficulties for the reductionistic Humean programme under discussion here. One might admit, for the sake of argument, that any causal path may be broken into chunks small enough to make the proposal work. But that is not enough. If the statistics in this kind of model are to double for causation, as the Hume programme supposes, some instructions must be given on how to construct the model from the kind of information about the situation that the programme can allow. But that cannot be done; and even to get a start in doing it will require the input of facts about individual causal histories.

The immediate lesson I want to draw from the fact that each of these strategies in turn fail, and in particular from the analysis of the way in which they fail, is apparent: there is no way to avoid putting in singular causal facts if there is to be any hope of establishing causal laws from probabilities. But in fact I think the discussion indicates a far stronger conclusion. Before I turn to this, there are two related loose ends that need some attention. The first involves the connection between the fact that a cause operates on a given occasion and the fact that it produces an effect on that occasion. The other concerns the connection between the propositional representation, using Boolean connectives, that follows Mackie's discussion, and the common representations of the social sciences, which use linear equations.

Throughout this chapter I have talked as if the operation of a probabilistic cause constituted a singular causal occurrence; and I have used the two interchangeably, assuming that an argument for one is an argument for the other. In the case of the linear equations this is true: whenever a cause operates, it succeeds in making its corresponding contribution to the effect. But where causes are all-or-nothing affairs, matters are somewhat different. What is supposed to occur when both a cause and a preventative operate? When effects can vary in their size, the two can both contribute their intended influences; if they are equal and opposite in size, they will leave the effect at its zero level. There is no such option in the propositional representation. On any given occasion, even if both the cause and the preventative operate, only one or the other of the two corresponding singular facts can occur. Either E results, in which case the cause produced it, or $\neg E$ results, having been caused by the preventative.

It does not seem that there should be a general answer to the question. Propositional models will be more appropriate to some situations than others; and what happens when warring causes operate will depend on the details of that situation, and especially on the relative strengths of the two opposing factors. Even in entirely qualitative models like Mackie's, where no question of the strength of a cause enters but only its occurrence is deemed relevant, different consistent assumptions can be made about how operations relate to singular causal truths. One among the many issues that must be faced in adopting one set of assumptions or another is the familiar question of over-determination: when two positive causes both operate, can it be true that they both produce the self-same effect, or must one win over the other? I do not have anything especially interesting to say about these questions. I see that in purely qualitative models, singular causings and operations may be distinct; that seems only to strengthen the argument against a Hume-type programme, since it will be necessary to admit both despised concepts. Rather than pursue these questions further, I shall instead focus primarily on the more common cases where causes and effects vary in their intensity, and where singular causings and operations converge.

3.5. Singular Causes In, Singular Causes Out

Not all associations, however regular, can double for causal connections. That is surely apparent from the discussions in this chapter. But what makes the difference? What is it about some relations that makes them appropriate for identifying causes, whereas others are not? The answer I want to give parallels the account I sketched in the introduction of physics' one-shot experiments. The aim there, I claim, is to isolate a single successful case of the causal process in question. Einstein and de Haas thought they had designed their experiment so that when they found the gyromagnetic ratio that they expected, they could be assured that the orbiting electrons were causing the magnetic moment which appeared in their oscillating iron bar. They were mistaken about how successful their design was; but that does not affect the logic of the matter. Had they calculated correctly all the other contributions made to the oscillation on some particular occasion, they would have learned that, on that occasion at least, the magnetic moment was not caused by circulating electrons.

In discussing the connection between causes and probabilities in Chapters 1 and 2, I have used the language of measuring and testing. The language originates from these physics examples. What Einstein and de Haas measured directly was the gyromagnetic ratio. But in the context of the assumptions they made about the design of their experiment, the observation of the gyromagnetic ratio constituted a measurement for the causal process as well. What I will argue in this section is that exactly the same is true when probabilities are used to measure causes: the relevant probabilistic relations for establishing causal laws are the relations that guarantee that a single successful instance of the law in question has occurred. The probabilities work as a measurement, just like the observation of the gyromagnetic moment in the Einstein–de Haas bar: when you get the right answer, you know the process in question has really taken place. That means that the individual processes come first. These are the facts we uncover using our empirical methods. Obviously, it takes some kind of generalization to get from the single causal fact to a causal law; and that is the subject of the next two chapters. Before that, I turn again to the points made in the earlier sections of this chapter and the last to support two claims: first, that the probabilities reveal the single case; and second, that the kinds of general conclusions that can be drawn from what they reveal will not be facts about regularities, but must be something quite different.

The philosophical literature of the past twenty years has offered a number of different relations which are meant to pick out causes or, in some cases, to pick out explanatory factors in some more general sense. When Hempel moved from laws of universal association, and the concomitant deductive account of explanation, to cases where probabilistic laws were at work, he proposed that the cause (or the *explanans*) should make the effect (the *explanandum*) highly probable.[29] Patrick Suppes amended that to a criterion closer to the social-science concept of correlation: the cause should make the effect more probable than it otherwise would be.[30] Wesley Salmon[31] argued that the requirement for increase in probability is too strong; a decrease will serve as well. Brian Skyrms[32] reverted to the demand for an increase in probability, but added the requirement that the probabilities in question must be resilient under change in circumstance. There are other proposals as well, like the one called 'CC' in Chapter 2 here.

What all these suggestions have in common is that they are non-contextual. They offer a single criterion for identifying a cause regardless of its mode of operation.[33] This makes it easier to overlook the question: what is so special about this particular probabilistic fact? One can take a kind of pragmatic view, perfectly congenial to the Humean, that causation is an empirical concept. We have discovered in our general experience of the world that this particular probabilistic relation is a quite useful one. It may, for instance, point to the kind of pattern in other probabilistic facts for other situations which is described in the next chapter. It turns out, one may argue, as a matter of fact, that this relation has a kind of predic-

[29] C. G. Hempel, *Philosophy of Natural Science* (Englewood Cliffs, NJ: Prentice-Hall, 1966).

[30] P. Suppes, *Probabilistic Theory of Causality* (Atlantic Highlands, NJ: Humanities Press, 1970).

[31] W. Salmon, *Statistical Explanation and Statistical Relevance* (Pittsburgh, Pa.: Pittsburgh University Press, 1971).

[32] B. Skyrms, *Causal Necessity* (New Haven, Conn.: Yale University Press, 1980).

[33] When CC is amended to CC^* this is, in effect, no longer true. For the suggestion of CC^* amounts to the proposal to hold fixed, not only all other causes, but all other operations as well. Of course, once the operations are admitted as essential, Hume's thesis about the primacy of the generic causal claim over the singular is already repudiated.

tive power beyond its own limited domain of applicability, and that others do not. This argument is harder to put forward once the context of operation is taken into account. One of the primary lessons of section 3.2 is that nature earmarks no single probabilistic relation as special. For instance, sometimes factorizability is a clue to the common cause, and sometimes it is not. This undermines the claim that it is the probabilities themselves that matter; and it makes it more evident that their importance depends instead on what they signify about something else—as I would have it, what they signify about the single occurrence of a given causal process.

Consider again the example of the birth-control pills. In section 3.3 I argued that the most general criteria for judging the causal efficacy of the pills would necessarily involve information about the degree of correlation between the occasions on which they operate to promote thrombosis and those on which they operate to inhibit it. The case where the operations are independent is a special one, but let us concentrate on it, since in that case the conventional methods of path analysis will be appropriate; and I want to make clear that even the conventional methods, when they work, do so because they guarantee that the appropriate singular process has occurred. The methods of path analysis are the surest of the usual non-experimental techniques for studying mixed capacities. Recall what these methods recommend for the birth-control example: to find the positive capacity of the pills to promote thrombosis, look in populations in which not only all the causal factors up to the time of the pills are held fixed, but so too is the subsequent state with respect to pregnancy; and conversely, to see whether they have a negative capacity to inhibit thrombosis, look in populations where, not P, but C' is held fixed. Why is this a sensible thing to do?

First consider the inference in the direction from probability to cause. Suppose that in a population where every woman is pregnant at t_2 regardless of whether she took the pills at t_1 or not, the frequency of thrombosis is higher among those who took the contraceptives than among those who did not. Why does that show that contraceptives cause thrombosis? The reasoning is by a kind of elimination; and it is apparent that a good number of background assumptions are required to make it work. Most of these are built into the path-theoretical model. For instance, it must be assumed that the probability for a spontaneous occurrence of thrombosis is the same in both groups, and also that the fixed combination of background causes will have the same tendency to produce thrombosis with C as without C. In that case, should there be more cases of thrombosis among those with C than among those without, there is no alternative left but to suppose that some of these cases were produced by C.

The same kind of strategy is used in the converse reasoning, from causes to probabilities. Assume that occurrences of C sometimes do produce thrombosis. The trick is to find some population in which this is bound to result in an increase in the probability of thrombosis among those who take the pills. Again, the reasoning is by elimination of all the alternative possibilities; and again, the inference makes some rather strong assumptions, primarily about the background rates and their constancy. If birth-control pills do sometimes produce thrombosis, there must be more cases of thrombosis when the pills are taken than when they are not—so long as the pill-taking is not correlated with any negative tendencies. But normally it is, since the pills themselves carry the capacity to prevent thrombosis. This is why P is held fixed. A pregnant woman who has taken the contraceptives is no more likely to be the locus of a preventative action from the pills than is a pregnant woman who has not. So the individual cases in which the pills cause thrombosis are bound to make the total incidence of thrombosis higher among those who have taken the pills than among those who have not. The same is true among women who have not become pregnant by t_2 as well. So holding fixed P is a good strategy for ensuring that the individual cases where the pills do cause thrombosis are not offset, when the numbers are totalled up, by cases in which they prevent it.

In both directions of inference the single case plays a key role: what is special about the population in which P is held fixed at t_2 is that in this population an increase in probability guarantees that, at least in some cases, the pills have caused thrombosis; and conversely, if there are cases where the pills do cause thrombosis, that is sure to result in an increased probability.

This is just one example, and it is an especially simple one since it involves only two distinct causal paths. But the structure of the reasoning is the same for more complicated cases. The task is always to find a special kind of population, one where individual occurrences of the process in question will make a predictable difference to the probabilities, and, conversely, where that probabilistic difference will show up only if some instances of the process occur. The probabilities serve as a measure of the single case.

The argument need not be left to the considerations of intuition. The techniques of path analysis are, after all, not a haphazard collection of useful practices and rules of thumb; they are, rather, techniques with a ground. Like the statistical measures of econometrics described in Chapter 1, they are grounded in the linear equations of a causal model; and one needs only to look at how the equations and the probabilities connect to see the centrality of the single case. Return to the discussion of causal models in Chapter 1. There, questions of causal inference were segmented into three chunks: first, the problem of estimating probabilities from observed data; second, the problem of how to use the probabilities to identify the equations of the model; and third, the problem of how to ensure that the equations can legitimately be given a causal interpretation. Without a solution to the third problem the whole program will make no sense. But it is at the second stage where the fine details of the connection are established; and here we must remind ourselves what exactly the probabilities are supposed to signify.

The probabilities identify the fixed parameters; and what in turn do these represent? Superficially, they represent the size of the influence that the cause contributes on each occasion of its occurrence. But we have seen that this is a concept ill-suited to cases where causes can work indeterministically. Following the lessons of section 3.3, we can go behind this surface interpretation: the fixed parameter is itself just a statistical average. The more fundamental concept is the operation of the cause; and it is facts about these operations that the standard probabilistic measures are suited to uncover: is the operation of the cause impossible, or does the cause sometimes contribute at least partially to the effect? The qualitative measures, such as increase in probability, or factorizability, are designed to answer just that question, and no more. More quantitative measures work to find out, not just whether the cause sometimes produces its putative effect, but how often, or with what strength. But the central thrust of the method is to find out about the individual operation—does it occur sometimes, or does it not? When the answer is yes, the causal law is established.

Recall the derivation of the factorizability criterion in Chapter 1, or the generalization of that criterion proposed in section 3.3. In both cases the point of the derivation is to establish a criterion that will tell whether the one effect ever operates to contribute to the other. Or consider again the path-analytical population used for testing whether birth-control pills cause thrombosis. In this case pregnancy is to be held fixed at some time t_2 after the ingestion of the pills. And why? Because one can prove that the probabilities for thrombosis among pill-takers will go up in this population if and only if $\hat{a}\hat{e} \neq \perp$; that is, if and only if the operation of the pills to produce C' sometimes occurs in concatenation with the operation of C' to produce thrombosis. Put in plain English, that is just to say that the probability of thrombosis with pill-taking will go up in this population only if pills sometimes do produce thrombosis there; and conversely, if the pills do ever produce thrombosis there, the probability will go up. What the probabilities serve to establish is the occurrence of the single case.

3.6. Conclusion

The first section of this chapter argued that singular causal input is necessary if probabilities are to imply causal conclusions. The reflections of the last section show that the output is singular as well. Yet the initial aim was not to establish singular causal claims, but rather to determine what causal laws are true. How do these two projects bear on each other? The view I shall urge in the next two chapters fashions a very close fit between the single case and the generic claim. I will argue that the metaphysics that underpins both our experimental and our probabilistic methods for establishing causes is a metaphysics of capacities. One factor does not produce the other haphazardly, or by chance; it will do so only if it has the capacity to do so. Generic causal laws record these capacities. To assert the causal law that aspirins relieve headaches is to claim that aspirins, by virtue of being aspirins, have the capacity to make headaches disappear. A single successful case is significant, since that guarantees that the causal factor does indeed have the capacity it needs to bring about the effect. That is why the generic claim and the singular claim are so close. Once the capacity exhibits itself, its existence can no longer be doubted.

It may help to contrast this view with another which, more in agreement with Hume, tries to associate probabilities directly with generic claims, with no singular claims intermediate. It is apparent from the discussions in this chapter that, where mixed capacities are involved, if causes are to be determined from population probabilities, a different population must be involved for each different capacity. Return to the standard path-analysis procedure, where the test for the positive capacity of birth-control pills is made by looking for an increase in probability in a population where pregnancy is held fixed; the test for the negative capacity looks for a decrease in probability in the population where the chemical C' is held fixed. This naturally suggests a simple device for associating generic claims directly with the probabilistic relations: relativize the generic claims to the populations.

Ellery Eells has recently argued in favour of a proposal like this.[34] According to Eells, a causal law does not express a simple two-placed relation between the cause and its effect, but instead properly contains a third place as well, for the population. In the case of the birth-control pills, for instance, Eells says that contraceptives are causally positive for thrombosis 'in the subpopulation of women who become pregnant (a few despite taking oral contraceptives)'.[35] They are also causally positive 'in the subpopulation of women who do not become pregnant (for whatever reason)'. In both these cases the probability of thrombosis with the pills is greater than the probability without. In the version of the example where the reverse is true in the total population of women, Eells concludes, 'in the whole population, [contraceptive-taking] is negative for [thrombosis]'.[36]

Under this proposal, the generic causal claims, relativized to populations, come into one-to-one correspondence with the central fact about association—whether the effect is more probable given the cause, or less probable. It should be noted that this does not mean that for Eells causal laws are reducible to probabilities, for only certain populations are appropriate for filling in the third place in a given law, and the determination of which populations those are depends on what other causal laws obtain. Essentially, Eells favours a choice like that proposed in Principle CC. The point is that he does

---

[34] E. Eells, 'Probabilistic Causal Levels', in Skyrms and Harper, op. cit., 79–97.
[35] Ibid.
[36] Ibid.

not use CC^*. For Eells, there is neither any singular causal input nor any singular causal output.

An immediate problem for Eells's proposal is that the causal laws it endorses seem to be wrong. Consider again the population of women who have not become pregnant by time t_2. On the proposal that lines up correlations and causes, only one generic relation between pills and thrombosis is possible, depending on whether the pills increase the probability of thrombosis or decrease it. For this case that is the law that says: 'In this population, pills cause thrombosis.' That much I agree with. But it is equally a consequence of the proposal that the law 'In this population, pills prevent thrombosis' is false; and that is surely a mistake. For as the case has been constructed, in this population most of the women will have been saved from thrombosis by the pills' action in preventing their pregnancy. What is true in this population, as in the population at large, is that pills both cause and prevent thrombosis.

Eells in fact substantially agrees with this view in his most recent work. His forthcoming book[37] provides a rich account of these singular processes and how they can be treated using single-case probabilities. The account is similar in many ways to that suggested in Principle*. Eells now wants to consider populations which are defined by what singular counterfactuals are true in them, about what would happen to an individual if the cause were to occur and what would happen if it were not to. The cause 'interacts' with the different sub-populations which are homogeneous with respect to the relevant counterfactuals. His notion of interaction seems to be the usual statistical one described on page 164; basically, the cause has different probabilistic consequences in one group than in another. It looks, then, as if we are converging on similar views.

That is Eells himself. I want to return to the earlier work, for it represents a substantial and well-argued point of view that may still seem tempting, and I want to be clear about what kinds of problems it runs into. In the earlier work of Eells all that is relevant at the generic level are the total numbers—does the probability of thrombosis go up or down, or does it stay the same? But to concentrate on the net outcome is to miss the fine structure. I do not mean by fine structure just that Eells's account leaves out the ornateness of detail that would come with the recounting of individual histories; but

---

[37] E. Eells, *Probabilistic Causality*, forthcoming.

rather that significant features of the nomological structure are omitted. For it is no accident that many individual tokens of taking contraceptives cause thrombosis; nor that many individual tokens prevent it. These facts are, in some way, a consequence or a manifestation of nature's causal laws.

Hume's own account has at least the advantage that it secures a connection between the individual process and the causal law which covers it, since the law is a part of what constitutes the individual process as a causal one. But this connection is hard to achieve once causes no longer necessitate their effects; and I think it will in fact be impossible so long as one is confined to some kind of a regularity view at the generic level, even a non-reductive one of the kind that Eells endorses. For regularity accounts cannot permit contrary laws in the same population. Whatever one picks as one's favourite statistic—be it a simple choice like Hempel's high probability, or a more complicated one like that proposed in Principle *CC*—either that statistic will be true of the population in question, or it will not; and the causal law will be decided one way or the other accordingly. This is just what we see happening in Eells's own account. Since in the total population the probability of thrombosis is supposed to be lower among those who take pills than among those who do not, Eells concludes, 'In the whole population [contraceptive-taking] is negative for [thrombosis]'; and it is not at the same time positive.

What, then, of the single cases? Some women in this population are saved from thrombosis by the pills; and those cases are all right. They occur in accord with a law that holds in that population. But what of the others? They are a sort of nomological dangler, with no law to cover them; and this despite the fact that we are convinced that these cases are no accident of circumstance, but occur systematically and predictably. Nor does it help to remark that there are other populations—like the sub-populations picked out by path analysis—where the favoured regularity is reversed and the contrary law obtains. For when the regularity is supposed to constitute the truth of a law, the regularity must obtain wherever the law does.

There are undoubtedly more complicated alternatives one can try. The point is that one needs an account of causal laws that simultaneously provides for a natural connection between the law and the individual case that it is supposed to cover, and also brings order into the methodology that we use to discover the laws. The account in terms of capacities does just that; and it does so by jettisoning regularities. The regularities are in no way ontologically fundamental. They are the consequence of the operation of capacities, and can be turned, when the circumstances are fortuitous, into a powerful epistemological tool. But a careful look at exactly how the tool works shows that it is fashioned to find out about the single case; and our account of what a law consists in must be tailored to make sense of that.

4

Capacities

4.1. Introduction

This book begins with a defence of causal laws, which have received rough treatment at the hands of a number of empiricist philosophers. Hume would reduce them all to associations; Mach and Russell would cast them out of science. But in the last fifteen years in philosophy of science, causes have become more acceptable. A number of authors already discussed—Clark Glymour, Elliott Sober and Ellery Eells, or Wesley Salmon—maintain that Russell and Mach were wrong. In addition to the concept of a functional relation or a probabilistic law, science needs a separate notion of causal law as well. I want to argue in this chapter that they have not gone far enough: in addition to the notion of causal law, we also need the concept of capacity; and just as causal laws are not reducible to functional and statistical laws, so too ascriptions of capacity are not reducible to causal laws. Something more is required.

Perhaps, though, that is a misleading way to put the point. For the concept of causation that is employed in most of these authors is already a concept of capacity; and I am very glad to have recognized this, since it brings my views more into line with those of others. For I maintain that the most general causal claims—like 'aspirins relieve headaches' or 'electromagnetic forces cause motions perpendicular to the line of action'—are best rendered as ascriptions of capacity. For example, aspirins—because of being aspirins—can cure headaches. The troublesome phrase 'because of being aspirins' is put there to indicate that the claim is meant to express a fact about properties and not about individuals: the property of being an aspirin carries with it the capacity to cure headaches. What the capacities of individuals are is another, very complex, matter. For instance, must the relevant conditions for the exercise of the capacity be at least physically accessible to the individual before we are willing to ascribe the capacity to it? These are questions I will have nothing to say about.

The claim I am going to develop in this chapter is that the concept of general *sui generis* causal truths—general causal truths not reducible to associations—separates naturally into two distinct concepts, one at a far higher level of generality than the other: at the lower we have the concept of a causal law; at the higher, the concept of capacity. I speak of levels of generality, but it would be more accurate to speak of levels of modality, and for all the conventional reasons: the claims at both levels are supposed to be universal in space and through time, they support counterfactuals, license inferences, and so forth. The level of capacities is 'higher' not only because the generalizations involved are broader in scope but also because the inferences that they allow are far stronger and, correlatively, the amount of metaphysics assumed in their use is far greater.

Part of my point is to make the metaphysics seem less noxious by making it seem more familiar. For I want to show that this higher-level concept is already implicit both in standard philosophical accounts of probabilistic causality and in the conventional methods for causal inference in medicine, agriculture, manufacturing, education, and the like—in short, anywhere that statistics are supposed to tell us about causes.

This point is important to the structure of my overall argument, for it ties together the work on causation at the beginning of the book with conclusions I draw here about capacities. I maintain that the crucial question for an empiricist must always be the question of testing. So we must ask: can capacities be measured? But that question has been answered, in Chapter 1 and in section 2.4. There I described how to use probabilities, treatment-and-control groups, and experiments constructed in the laboratory to test causal claims. What I want to show here is that these causal claims are claims about capacities, and that we already know how to test them.

4.2. Why Should Increases in Probability Recur?

If one looks in the recent philosophical literature on probabilistic causality, a number of works begin with roughly the same formula—the formula called Principle CC in Chapter 2. The formula says that 'C causes E' if and only if the probability of E is greater with C than without C in every causality homogeneous background. It must be apparent from the discussions of Chapters 2 and

3 that the exact formulation of the principle is a matter of contro-
versy; and there has been a good deal of discussion of how the
formula errs, what kinds of caveat need to be added, and so forth. I
want to focus in this section on an aspect of the formula that has
received little attention. The formula as I have written it in Chapter 2
appears with a strict inequality. Brian Skyrms[1] writes it instead with
the left-hand side 'greater than or equal to' the right-hand side.
There has been discussion of that issue. But there has been little
notice of what seems to me the most striking feature of this
formula—the universal quantifier in front: C is to increase (or at
least not decrease) the probability of E in *every* homogeneous background. My central thesis here is that the concept of causality that
uses such a universal quantifier is a strong one indeed; strong enough
to double as a concept of capacity.

Some attention, of course, has been devoted to the matter of the
universal quantifier in Principle CC. The last chapter was full of
such discussion; and John Dupré, who has been deeply involved in
the issues discussed there, even has a name for the phenomenon,
which I shall use too: causes, he says, have *contextual unanimity*
when they change the probability in the same direction in every
homogeneous background. If the probability of the effect sometimes goes up with the presence of the cause, and sometimes goes
down, then there is a lack of contextual unanimity. The reason I
claim the quantifier is not much noticed, despite the discussions of
the last chapter, is that the discussion there, like the discussion of
Eells and Sober, already presupposes the conceptual framework
of capacities. In Hesslow's example, birth-control pills have, I
claimed, *dual capacities*; and their probabilities exhibit the same
kind of systematic behaviour which, I will claim, is characteristic of
situations where capacities are at work. In a sense, to start from
failures of unanimity of the kind we see in the contraceptives
example would be to beg the question. I would prefer, then, for the
nonce, to disallow the possibility of dual capacities, and to ask the
antecedent question: why expect any pattern of probability relations
from one background setting to another, let alone the totally uniform pattern implied by the universal quantifier? By focusing only
on examples where capacities are not dual, we will be able to sort out
cases where the universal quantifier fails in a systematic way and for

[1] *Causal Necessity* (New Haven, Conn.: Yale University Press, 1980).

good reason—because an opposing capacity is at work—from cases where no pattern at all is to be expected, because no capacity exists.

It will help to return to the major conclusions of the last chapter. I of course endorse CC^*, and not the simpler CC which Eells and Sober opt for, and probably Skyrms as well. In either case, one can ask: what is supposed to follow from an increase in probability of E on C in one of the specially selected populations (which I will henceforth call 'test situations')? According to Chapter 3, what follows is that, in those special test populations, it can be regularly relied on that some Cs will produce Es. The argument has two parts. One is already presupposed by the reference to probabilities rather than to actual frequencies. The use of the probability concept guarantees that there will be reliably (on average or in the long run) more Es when C is present than when C is absent. The further conclusion that at least some of the Es must be produced by Cs depends on the specific features these test populations are suppposed to have. Roughly, enough is supposed to be true about them to guarantee that if there is an increase in the number of Es, there is no account possible except that C produced the excess. The point I want to make about this argument is that it justifies a very local kind of causal claim: if in a given test population we see the increase in probability that we are looking for, that guarantees that Cs cause Es *there in that population*. But it does not tell us any more. Since it is probabilities and not mere frequencies that are involved, it is assured that the causing of Es by Cs will happen regularly—but regularly in that kind of population. Who knows what happens elsewhere?

From now on I will call these very local kinds of causal claim 'causal laws', and I shall adopt the proposal of Ellery Eells, mentioned in the last chapter, that causal laws—in this sense—have, not two, but three places in them: causal laws are relative to a particular test population. My objection in the last chapter was not to the relativization itself. It seems to me that the kind of causal claim which is immediately supported by finding an increase in probability must be restricted to the population in which the increase occurs; even then it is only supported if that population has the very special characteristics that permit an inference from correlations to causes. What I objected to in the last chapter was the claim that the increase in probability in one of these special populations constitutes the truth of the causal law there. What makes the causal law true that C causes E in T is not the increase in probability of E with C in T, but rather the fact that in *T* some *C*s do regularly cause *E*s. The increase in probability is only a sign of that; and, as I argued in the last chapter, there will generally be other causal laws true as well in the same population—laws which are not revealed in the probabilities.

If we adopt the convention of calling these local truths 'causal laws', we must look for another term to express the more usual concept of causation that philosophers have been grasping for in their various probabilistic theories of causality. For causal laws are now to be relativized to particular test situations. Yet the initial formula—either *CC* or *CC**—quantifies over *all* test situations. What kind of a concept of causation does this universal quantification represent? Obviously a concept stronger and more general than the (relativized) causal laws just introduced. Of course one can de-relativize: do not consider '*C*s cause *E*s in *T*', but rather '*C*s cause *E*s', *simpliciter*. But what does this de-relativized claim say? There is a specific answer to that for the relativized version, an answer already apparent from the argument above. If the relativized law is true, then in *T* some *C*s will produce *E*s, at least on average, or in the long run, or however one wants to read probability. What about the non-relative version? I think it reports that *C*s—by virtue of being *C*—can cause *E*. In short, *C* carries the capacity to produce *E*. I think that in part because the universal quantifier involves just the kind of reasoning that is appropriate to capacities.

Consider why anyone would put that quantifier in. Recall that not many people note the quantifier. I do not know explicitly how it got there in other people's work. But I can explain why I have used it in introducing my version of either *CC* or *CC**. I have always used it specifically because I think in terms of causal capacities. If *C*s do ever succeed in causing *E*s (by virtue of being *C*), it must be because they have the capacity to do so. That capacity is something they can be expected to carry with them from situation to situation. So if the probability goes up in one test situation, thus witnessing to the capacity, it will do the same in all the others. Hence the universal quantifier. Recall that the assumption that the probabilities will go the same direction in all test situations[2] has a name: it is the assumption of contextual unanimity. So my thesis in short is this: to believe in contextual unanimity is to believe in capacities, or at least it is a good way along the road to that belief.

---

[2] Barring interaction, which will be discussed later.

It is important to remember that the use of this universal quantifier is widespread. That is, philosophers trying hard to characterize a concept of causal law commonly supposed that the concept they were aiming for guaranteed contextual unanimity. Yet most would eschew capacities on Humean grounds. There is more to capacities than contextual unanimity, I will agree. The more may be more palatable, though, if one only realizes how much has already been admitted in assuming that, where causes operate, there will be contextual unanimity. For contextual unanimity is a very peculiar concept for a Humean.

To see this, compare another now deeply entrenched concept, but one equally objectionable from an empiricist point of view—the concept of a law of nature. With all the work on possible-world semantics, philosophers seem to have become used to laws of nature, and the counterfactuals they license. So it is important to keep in mind how deeply this concept offends the basic tenets of a radical empiricist. Consider the very simple diagramatic example: all *A*s, by law, are *B*s. This law allows us to infer from one empirical fact, that something is an *A*, to another, that it is a *B*; we are able to know this second empirical fact without anybody ever having to look. This way of putting it can be misleading, though, for the matter is not just epistemological but metaphysical as well. When a law of nature is assumed, one empirical fact can constrain, or determine, the occurrence of another. But it is just these kinds of internal connection among empirical properties that empiricists set out to abolish.

The point of this short digression is that capacities are peculiar, from a radical empiricist point of view, in exactly the same way as laws, only one level up. Just as laws constrain relations between matters of fact, capacities constrain relations among laws. A property carries its capacities with it, from situation to situation. That means that, where capacities are at work, one can reason as above: one can infer from one causal law directly to another, without ever having to do more tests. In this way capacities are much like essences. If you are committed to the assumption that all the internal properties of electrons are essential, this makes science a lot easier for you. You can measure the charge or mass on one, and you know it on all the others. What I have been trying to show here is that it is a concept with just this peculiar kind of strength that is marked out by the universal quantifier in Principle *CC*.

In the next section, I want to show that it is a concept with exactly this same kind of peculiar modal strength that is presupposed by the conventional statistical methods used in the behavioural and life sciences. But first I summarize the three points that I hope have been established so far: first, that standard philosophical accounts of probabilistic causality employ a concept of causation stronger than the already troublesome concept of causal law; second, that the features that make it stronger are just those one would expect from a concept of capacity; and third, that if you are a radical empiricist, you should be very troubled by this concept, because the features that make it stronger introduce a peculiar modal force that no empiricist should countenance.

So far I have focused on modality as an inference licence. Yet in the Introduction I said that ascriptions of capacity are at a higher level of modality than causal laws not only because they license stronger inferences but also because they make stronger presuppositions about the stability and regularity of nature. The two are intimately connected, and both are closely related to my doctrines about testability and empiricism. Consider the question of induction and its scope. My arguments for a distinction between causal laws and ascriptions of capacity rest in part on claims about how general causal facts are established. Causal laws, which are relativized to certain given kinds of situation, can be inferred from the probabilities that obtain in those situations. To infer the stronger claim—what I call a capacity claim—one must suppose that the causal possibilities that are established in that situation continue to obtain in various different kinds of situation. It may seem that this makes for no great difference between the two kinds of causal claim: induction must be used in any case; what differentiates the two is just that an induction of wider scope is required for the stronger claim.

I think this suggestion is a mistake. On my view, induction never enters. The logic of testing is a bootstrap logic: the hypothesis under test is to be deduced from the data plus the background assumptions. Although inspired by Clark Glymour's account, this description is different from Glymour's own. I insist that in a reliable test the hypothesis itself should be derivable; Glymour asks only for an instance of the hypothesis. Then he still has to confront questions of the scope of the induction from the instance to the final hypothesis.

Of course there is ultimately no way to avoid these questions. In a sense, answers to them must already be built into the background assumptions if my version of bootstrapping is to succeed. But they are not always built in directly as explicit assumptions about scope. That is what much of this book has been about—how to deduce causal laws from probabilities—and no mention of scope has been made. If the generalized Reichenbach Principle is accepted, Chapter 1 shows how causal laws may be derived from facts about conditional probabilities—is the conditional probability of the effect with the putative cause present greater than, equal to, or less than the conditional probability with the cause absent? Scope enters only at the next stage. For the conditional probabilities are partial; other factors must be held fixed. The arguments of Chapter 1 allow causal laws to be established only one by one, population by population. Each different arrangement (or different level) of the other factors constitutes a new situation where the conditional probabilities must be assessed anew before a causal conclusion can be drawn.

One may of course be prepared to infer from the facts about conditional probabilities under one arrangement to facts about conditional probabilities in another. But that inference requires a different kind of licence, not yet issued; and in the bootstrap methodology that licence is provided by a correspondingly strong assumption in the background metaphysics—this time an assumption explicitly about the scope of induction, or the extent of regularity in nature. The same is true for the arguments earlier in this section, as well as in Chapter 2. Whether one follows the strategy of Chapter 1, 2, or 3, in no case can a causal law be admitted unless it has been tested. In all cases there must be a deductive argument that takes you from the data plus the background assumptions to the causal conclusion; and in all cases the arguments for the more specialized, lower-level claims, which I am now calling causal laws, require substantively weaker premises than do the arguments for the more general claims usually made in science, which I render as ascriptions of capacity.

4.3. Forecasting and the Stability of Capacities

I turn now to the linear equations, common to causal modelling and to path analysis, that provide a rigorous ground for standard probabilistic measures of causality. I am going to discuss econometrics, and indeed very early econometrics, because my points are particularly patent there. But my conclusions should apply wherever these conventional probabilistic methods are put to use. Recall that econometrics got under way at the end of the 1930s and in the 1940s with the work of Jan Tinbergen, Ragnar Frisch, and Trygve Haavelmo. The work that these three began, along with Tjalling Koopmans, was carried out and developed at the Cowles Commission for Economic Research in Chicago. I give a quite distilled version of what Haavelmo and Frisch were doing. In order to keep the story all together, I will begin again at the beginning, and not rely too heavily on the various discussions of econometrics from earlier chapters.

Consider first a simple linear example familiar to everyone: a price–demand curve:

$$D$$
$$q = \alpha p + u$$

q represents quantity demanded, and p, price; u is supposed to represent some kind of random shock which turns the equation from a deterministic relationship into a probabilistic one. It is the fixed parameter α that is my focus of interest, and I will come to it in a moment.

First, though, it is important to note that this equation is supposed to represent a causal relationship, and not a mere functional relationship of the kind that Russell thinks typifies physics. Econometrics arises in an economic tradition that assumes that economic theory studies the relations between causes and effects. This assumption is so basic that, in a sense, it goes without mention. Few economists say, 'We are talking about causes'. Rather, this is a background that becomes apparent when topics which are more controversial or more original are discussed. I note just one example.

Jakob Marschak was head of the Cowles Commission during the important years (1943–8) when the basic ideas were worked out for what are now called Cowles Commission methods. In 1943, at the end of his first year as director of research, Marschak describes the fundamental ideas of three studies begun that year.

The method of the studies . . . is conditioned by the following four characteristics of economic data and economic theory: (a) the theory is a system of simultaneous equations, not a single equation; (b) some or all of these equations include 'random' terms, reflecting the influence of numerous erratic causes in addition to the few 'systematic' ones; (c) many data are given in the form of time series, subsequent events being dependent on preceding ones; (d) many published data refer to aggregates rather than to single individuals. The statistical tools developed for application in the older empirical sciences are not always adequate to meet all these conditions, and much new mathematical work is needed.[3]

What I want to highlight is the language of the second characteristic: the random terms reflect the *influences* of erratic causes, which operate in addition to the few systematic ones. So, as I said, the methods aim to study the relationship between causes and effects; and the simple demand equation, *D*, follows the standard convention of writing causes on the right and effects on the left. Here the effect is quantity demanded, which is supposed to be determined simultaneously by the systematic influence of the price plus a number of other random erratic factors.

But there is more contained in this equation, more to displease Hume or Russell, than just the assumption of causality. The equation does not just assume that from one occasion to another the price causes, or is a contributing cause to, the demand, in some haphazard or unsystematic way. Rather, it assumes that the price has a stable tendency to influence demand, and that that tendency has a fixed and measurable strength. That is what is represented by the α, and by the way it is treated.

Consider α then: α represents the price elasticity of demand, a concept that was significant in the marginal revolution. But to my mind, α is treated very differently by the econometricians from the way it was treated, for example, by Alfred Marshall. When Marshall introduced the idea of demand elasticity in his *Principles of Economics*, he immediately proceeded to consider 'the general law of the *variation* of the elasticity of demand'. How does the elasticity *change*? For instance, the price elasticity of demand will itself depend on the *level* of the price: the elasticity of demand is great for high prices, and great, or at least considerable, for medium prices; but it declines as the price falls; and gradually fades away 'if the fall goes so far that the satiety level is reached'.[4] In general, the econometricians by contrast treated the elasticity as if it did not vary: α

[3] C.F. Christ, 'A History of the Cowles Commission, 1932–1952', in Cowles Commission for Research in Economics, *Economic Theory and Measurement: A Twenty Year Research Report, 1932–1935* (Chicago, Ill.: Cowles Commission, 1952), 31.

[4] A. Marshall, *Principles of Economics* (London: Macmillan, 1907), 103.

measures an abiding or stable tendency of the system. Of course the two views are reconcilable, as is even suggested by the terminology. For the elasticity of a material is a characteristic of the material that remains fairly fixed over a wide range of conditions to which the material is normally subject; yet it can shift quite dramatically outside this range.

What I want to stress is that the assumption of stability is built into the demand equation. From the discussion of Marshall one would expect that α was a function of p, i.e. $\alpha = \alpha(p)$, and of a number of other features as well: $\alpha = \alpha(p, . . .)$. But that is not how it is treated. It is treated as a constant—or fixed—parameter. Of course, econometricians do sometimes model price-dependent elasticities. I am using this simplified example to show what assumptions are hidden in the conventional methods. Let me give an example to illustrate the particular kind of stability assumption that I think is pre-supposed in these methods.

We have seen that, owing to the presence of the u, the demand equation does not represent a deterministic relationship between p and q. Instead, there will be some probability distribution for the two. Assume, as is usual, that it is a bi-variate normal distribution. Recall from Chapter 2 that a bi-variate normal distribution can be completely identified by specifying five parameters: $\mu_p, \mu_g, \sigma_p, \sigma_g, \rho$, where it is conventional to scale the quantities so that both the means are zero ($\mu_p = 0 = \mu_q$) and both variances are 1 ($\sigma_q = 1 = \sigma_p$). Then the conditional distribution of q for any fixed value of p will again be normal, with mean ρp; and q can always be written

$$q = \rho p + u$$

which (so long as u has mean 0 and variance $1 - \rho^2$) is just the demand equation, if α is set equal to ρ.

The point of this is that you can go back and forth between the conditional distribution and the demand equation. The equation summarizes the same information as the distribution, and vice versa. So what is the point of the equation? This is an old question in causal-modelling theory; and there is a conventional answer. In the words of Marschak himself, 'the answer to this question involves . . . [a] discussion of the degree of permanence of economic laws.'[5]

[5] W.C. Hood and T.C. Koopmans, *Studies in Econometric Method*, Cowles Commission Monograph 14 (New York: Wiley, 1953), ch. 1, s. 3.

Or, as O. Duncan in his classic *Introduction to Structural Equation Models* says,

there would not be much purpose in devising a model . . . if we did not have some hope that at least some features of our model would be invariant with respect to some changes in the circumstances. . . . If all the model is good for is to describe a particular set of data . . . then we might as well forego the effort of devising the model. . . .[6]

The point of the model, then, is that its equations express a commitment about what remains constant under change. In the case of the demand equation, that is supposed to be α. The point of expressing the information about a particular bi-variate normal distribution in the form of an equation is to signal the commitment that it is the ratio which stays fixed no matter how the variances shift. It follows from formula D at the beginning of this section that $\alpha = \mathrm{Exp}(pq)/\mathrm{Exp}(p^2) = \rho/\sigma_p^2$. What happens when the variance of p changes? Now we will have a new distribution in p and q; even if we assume that it too is bi-variate normal, the parameters for the distribution will still be unsettled. Nature might, for instance, be very concerned with correlations, and fix it so that $\mathrm{Exp}(pq)$ stays fixed no matter what the variance in p is. Or it could be instead that $\mathrm{Exp}(pq)$ is always varied in just the right way to keep α fixed. And so on.

Consider a different change. Imagine that we wish to adapt our study to more modern times, including the amount of television advertising as an independent cause that can also exert an influence on quantity demanded. Call this new variable r. The natural way to modify the demand equation is this:

$$D'$$

$$q = \alpha p + \beta r + u$$

Just add r as an additional influence, but assume that α stays the same. That is, *whether r is there or not*, we assume that the price elasticity stays the same.

It is important to see how peculiar this is. From the point of view of the new equation, the old equation expresses a relationship that holds in one particular circumstance: $r = 0$. When r is 0, p and q

[6] O. Duncan, *Introduction to Structural Equation Models* (New York: Academic Press, 1975).

have a familiar distribution: a bi-variate normal. In principle, the distribution of p and q could be entirely different when r takes on some different value. But the way the equation is written denies that. The method of writing the equation assumes that, however p affects q, that is something that stays the same as the situation with respect to r changes, however it does so.

That, in general, is just what is reflected in econometric method. Parameters are estimated in one context, and those values are assumed to obtain in entirely different contexts. Couple that with the observation I made earlier that these parameters connect causes and effects, and you see why I talk here about stable causal tendencies. The methods presuppose that causes have stable tendencies of fixed strengths that they carry about with them from situation to situation. What p contributes to q—its total influence, αp—depends on p alone, and is the same no matter what goes on elsewhere. This is, in effect, the consequence of the commitment to the existence of a single linear theory that is supposed to hold across all different contexts and across all different statistical regimes.

Both the adjectives *single* and *linear* are important. It is not just the fact that one general theory is assumed across a variety of specific cases that brings in capacities, but rather the special form of that theory. Constraints on forms of theory are familiar, but the particular constraint that the theory be given in a set of linear equations, where the parameters are fixed, is an exceedingly strong one. Compare, for instance, the restriction that one might want to place on a modern quantum field theory, that it be renormalizable. Although this constraint is very limiting, it does not have the very peculiar feature that the assumption of the stability of the parameters produces for a linear model: namely that, given the constraint on the form, the theory itself can be inferred from the data. Of course, in econometrics the theory is not constructed from the 'raw' data, of observed frequencies in finite populations, but rather from the 'interpreted' data, or probabilities, that these frequencies are used to estimate. But this should not obscure the fact that the form allows us to discover in a single context the exact content of the theory that will cover all the rest.

It is partly because of this peculiarity that econometricians do not hold too uncritically to their commitment to a linear form with fixed parameters. In particular, they are always trying to determine, usually case by case, what kinds of variation in the parameters will provide better predictions. Indeed, the entire question is a matter of much discussion in econometrics at the moment, though the language used there is different from mine. Econometricians nowadays talk about targets, manipulations, and forecasting.[7] Their fundamental concern is: will the parameters that have been estimated, under whatever conditions obtained in the past, continue to hold under various innovations? If not, their models will be of no help in forecasting the results that would occur should the economy be manipulated in various proposed ways.

One good example can be found in a well-known critique of the use of econometrics in policy evaluation, by Robert Lucas.[8] Lucas argues that, in general, the parameters estimated in econometrics will not be stable across policy shifts, although they may be of use for forecasting where no policy change is envisioned. The primary reason is that the equations of econometrics, for which parameters are estimated, describe connections at the wrong level: they connect some 'forcing variables', which policy-makers might hope to influence, with macroscopic and observable states of the economy. The parameters are estimated from past observations. The policy-maker hopes to set the exogenous, or forcing, variables at some level, and expects that the resulting state of the economy will be that predicted by the equation. But that will not work, Lucas argues, because in general the parameters themselves will not be independent of the level at which policy sets the exogenous variables. That is because the macroscopic equations derive from some more fundamental equations which describe the decision procedures of individuals. But the structure of these more fundamental equations will be affected by the agents' expectations about policy shifts, and so the stability of the macroscopic parameters across policy changes is undermined: the macroscopic parameters are fixed by those at the micro-level; but these in turn respond to expectations about the level of the macroscopic forcing variables. So it is not likely that the parameters estimated under different policies will continue to be appropriate should policy be shifted. In Lucas's own words:

To assume stability of [the macroscopic equations and parameters] under alternative policy rules is thus to assume that agents' views about the

[7] R.F. Engle, D. Hendry, and J.F. Richard call parameters which have the necessary invariance to sustain policy predictions 'super-exogenous'. See 'Exogeneity', *Econometrica*, 51 (1983), 277–304.

[8] 'Econometric Policy Evaluation: A Critique', *Carnegie–Rochester Conference Series on Public Policy: The Phillips Curve and Labor Markets*, i (1976), 19–46.

behavior of shocks to the system are invariant under changes in the true behavior of these shocks. Without this extreme assumption, the kinds of policy simulations called for by the theory of economic policy are meaningless.[9]

Is Lucas's pessimism justified? This has been a matter of some debate. I bring up the example neither to endorse nor to reject his conclusions, but merely to stress how aware econometricians now are that the success of their methods depends on the kind of stability which I have been associating with the concept of capacity.

This was also a matter that was clear to the founders of econometrics, although it has not been much discussed between then and now. Lucas himself remarks that the criticisms he raises 'have, for the most part, been anticipated by the major original contributors'[10] to econometrics. He cites Marschak and Tinbergen. I want instead to turn to Frisch and Haavelmo because of the richness of their philosophical views. A central concept behind their work, in both cases, was that of *autonomy*. The difference between autonomous and non-autonomous laws is like the difference between fundamental laws, which hold by themselves, and derived or conditional laws, which hold on account of some particular (non-necessary) arrangement of circumstances. Haavelmo's own illustration explains it well:

Here is where the problem of *autonomy* of an economic relation comes in. The meaning of this notion, and its importance, can, I think, be rather well illustrated by the following mechanical analogy:

If we should make a series of speed tests with an automobile, driving on a flat, dry road, we might be able to establish a very accurate functional relationship between the pressure on the gas throttle (or the distance of the gas pedal from the bottom of the car) and the corresponding maximum speed of the car. And the knowledge of this relationship might be sufficient to operate the car at a prescribed speed. But if a man did not know anything about automobiles, and he wanted to understand how they work, we should not advise him to spend time and effort in measuring a relationship like that. Why? Because (1) such a relation leaves the whole inner mechanism of a car in complete mystery, and (2) such a relation might break down at any time, as soon as there is some disorder or change in any working part of the car. . . . We say that such a relation has very little *autonomy*, because its existence depends upon the simultaneous fulfillment of a great many other relations, some of which are of a transitory nature.[11]

[9] Ibid. 25.
[10] Ibid. 20.
[11] T. Haavelmo, *The Probability Approach in Econometrics*, supplement to *Econometrica*, 12 (July 1944), 27, 28.

Both Haavelmo and Frisch agreed that in non-autonomous laws the parameters might not be independent of each other, or even of the level of the other variables. Imagine, say, two variables x_1 and x_2 in a long equation. If that equation derives from some others, then the parameters for these two variables could readily be functions of some more fundamental ones. So manipulations to change α_2 may well result in variation in α_1. But what is more striking, and why these two econometricians are such good illustrations for me, is that both assumed that in the truly fundamental equations the parameters would be independent; that means that they really assumed that the fundamental structure of nature was one where causes could be assigned stable capacities of fixed strengths.

This was recognized by the opposition as well. As I have remarked before, one of the most notable critics of econometrics at its start was John Maynard Keynes. Keynes saw that econometrics did make just this assumption, and that was one of his principal objections to it. Nature, thought Keynes, did not work that way. According to Keynes, the methods of econometrics assumed that

The system of the material universe must consist . . . of bodies which we may term (without any implication as to their size being conveyed thereby) *legal* atoms, such that each of them exercises its own separate, independent, and invariable effect, a change of the total state being compounded of a number of separate changes each of which is solely due to a separate portion of the preceding state. We do not have an invariable relation between particular bodies, but nevertheless each has on the others its own separate and invariable effect, which does not change with changing circumstances, although, of course, the total effect may be changed to almost any extent if all the other accompanying causes are different. Each atom can, according to this theory, be treated as a separate cause and does not enter into different organic combinations in each of which it is regulated by different laws.[12]

With respect to the central issue here—the question of capacities—it is worth noting that Keynes's criticism of econometrics is, not surprisingly, more radical than the Lucas critique. I say this is not surprising because Lucas's specific arguments are based on reflections on the relationship between the more autonomous laws that govern behaviour of individual firms and individual consumers, on the one hand, and the less autonomous, aggregative laws of macro-economics, on the other. Enduring capacities disappear from

---

[12] J. M. Keynes, *A Treatise on Probability*, (London: Macmillan, 1957), 249.

the macro-level, where parameters may drift in ways which seem unintelligible given the information available at that level, only to reappear at the micro-level. There they are not, of course, immutable, for they can at least be affected by expectations about future policies—that, after all, is the core of Lucas's argument. It is also an important feature of the philosophical story, for it fits the capacities into nature in a realistic way. They do indeed endure; on the other hand, their characteristics may evolve naturally through time, and they may be changed in systematic, even predictable, ways as a consequence of other factors in nature with which they interact. All this speaks in favour of their reality.

Keynes proposes a very different kind of world: not an atomistic but a wholistic world in which the behaviour of any particular feature—the very contribution it makes to the effect from one occasion to another—depends on the setting in which it is embedded. To return to the main theme of this chapter—the two levels of causality concept—this would not be a world in which capacities operate; but it might well be, and for Keynes probably would be, a world where causal laws are at work. For causal laws are allowed to be context-dependent. What is necessary for the truth of the law 'C causes E in T', as characterized in section 4.2, is that in T, Cs can be regularly relied on to produce or contribute to the production of Es. Sometimes this can be discovered by what might be characterized as the 'autopsy method'—i.e. tracing the process by which C produces E in specific individuals. But that is not a method easy to employ in economics, and I have not discussed it much here. More typical would be to use statistics. If T is the right kind of situation—it has the characteristics of a 'test' situation—then a simple increase in probability of E on C will serve to establish the law. In the far more typical case where the situation of interest does not have the tidy properties that make for a test, one must consider how to construct some special test situation consistent with it. But, in any case, only the causal law will be established, and if the commitment to capacities is missing, no further inferences can be drawn. One could make an econometric model, but the enterprise would be otiose. For the model would offer no generalization over the facts already known. This is why I say, echoing Duncan, that the point of making the model is to endorse certain predictions and certain plans; and that endorsement only makes sense where an appropriate degree of autonomy can be presupposed.

Econometrics is a notoriously uncertain science, with a spotty record for predictive success. Perhaps that is because Keynes is right; but perhaps it comes about for any number of less deeply metaphysical reasons. One principal claim I make here is, not that the phenomena of economic life are governed by capacities, but rather that the methods of econometrics presuppose this, at least if they are to be put to use for predicting and planning. But the claim is more far-reaching than that. I talk a great deal about econometrics, but econometrics serves just as an exemplar where the methods are laid out in a particularly rigorous way, and where the metaphysics that makes these methods reasonable is particularly transparent. What I claim for econometrics will be equally true in any field that uses conventional statistical methods to make inferences about causes. For it has been my thesis throughout that it takes something like the structural equations of econometrics (or path analysis) to secure a reliable connection between causes and probabilities; and it is just the feature of stability that I have pointed to in econometric structures that must be presupposed in any discipline if we are to infer from the statistics we observe in one set of conditions to the effects that might be achieved by varying these conditions. In the next chapter I will argue that the same is true of the methods of physics; and, whatever is the case in economics, in physics these methods work, and the capacities that justify them are scarcely to be rejected. Before that, I want to explain why capacity claims should not be thought of as just higher levels of modality, but instead must be taken as ascriptions of something real.

4.4. Beyond Modality

The last two sections argued that contemporary behavioural and life sciences, like sociology, medicine, and econometrics, do a variety of things that require a concept of causation far stronger than that of causal law; and that these are just the kinds of thing that one would expect to be able to do with a concept of capacity. The next chapter will argue that this is true of physics as well. Still, this does not force one to admit capacities as real. After all, the argument so far only shows that a level of modality, or of generalization, is required beyond the level already needed for causal laws. In a sense I have done a disservice to my own point of view. I aimed to defend capacities as real; what I have shown is how to modalize them away. But that is not the last word. For the concept of capacity involves more than just ascending levels of modality. This section will explain what more there is.

First, to understand what I mean by 'modalizing capacities away', consider how a logical positivist might respond to the arguments of the last two sections. I think that positivists should welcome them. Rudolf Carnap is a good case.[13] Carnap assumed that causality is a concept used to say things in the material mode that would be more perspicuous if put in the formal mode. That is, when you use the concept of causality, you sound as though you are talking about the world, but in fact you are talking about linguistic representations of it. Bas van Fraassen[14] has more recently argued the same view, but with a semantic idea of representation in contrast to Carnap's syntactic one. Sections 4.2 and 4.3 above are squarely in this positivist tradition: they fill in content for the structure that both van Fraassen and Carnap prescribe. In the picture provided so far, claims about capacities do not seem to report facts about powers, dispositions, or activities. Nor do they report causal laws. Rather, they function as metalinguistic summaries of facts about causal laws; and this is just like Carnap's original idea about causal laws themselves, only one modal level higher. To carry on the positivist program, the causal laws themselves must be recast in the formal mode as summaries of facts about (or sets of contraints on) non-causal laws—functional relations and probabilities. These laws too are to be eliminated in turn, in favour of merely occurrent regularities. One by one each level of modality is to be stripped from the material mode and relocated in the formal. The levels are pictured in Fig. 4.1.

Why won't this strategy work? Consider first the relation between the lowest modal level, that at which natural laws occur, but only laws of association, and the next level up, where causal laws are admitted as well. The answer to the question, for these two levels, is already implicit in Chapter 3. To see how it works, it is first necessary to reconstruct the earlier discussions of the connection between causal laws and probabilistic laws in order to highlight the modal aspect of this connection. Return to Principle *CC*. Although it was not

[13] R. Carnap, *The Logical Structure of the World* (Berkeley, Calif.: University of California Press, 1967).
[14] *The Scientific Image* (Oxford: Clarendon Press, 1980).

Levels of modality:	Ascriptions of capacity Causal laws Functional and probabilistic laws
Non-modal level:	Occurrent regularities

FIG. 4.1

described this way in the original discussion in Chapter 2, whencast into the framework of Carnap or van Fraassen, Principle *CC* functions as a modal principle, one rank above the level of the laws of association. It is one modal level higher for exactly the same reasons that capacity claims are modal relative to lower-level claims about causal laws. The claims at the higher level constrain what structure the total set of facts at the lower level can have, and thereby license inferences from one kind of fact at the lower level directly to another, without the need for any support from below.

Roughly the constraints work like this:[15] a cause is supposed to increase the probability of its effect, holding fixed all the other causes. This will put a complicated restriction on the probability structure. Imagine, for example, that a new factor, C_n, raises the probability of E when $C_1, \ldots, C_{n-1}$ are held fixed. This means that C_n should be admitted as a cause of E, relative to $C_1, \ldots, C_{n-1}$. But now, what about each of these factors themselves? We may imagine that up until this point they have been correctly counted as causes, so that each raises the probability of E relative to the others. But now that C_n has been added, the judgement about each of $C_1, \ldots, C_{n-1}$ must be re-evaluated. If C_n is to be a cause, then each of $C_1, \ldots, C_{n-1}$ must increase the probability of E when C is held fixed as well. If one of these factors fails to do so, real complications follow. For in that case, it ought not to have been held fixed in the first place, in the evaluation of C_n itself. But what then about C_n? The process of evaluation must begin again; and sometimes, for some probabilistic structures, there will be no choice of a set of causal factors possible for a given effect, consistent with Principle *CC*.

This means that *CC* can be used to rule out some probability struc-

[15] For more details see 'How Some Worlds Could Not be Hume Worlds', in N. Cartwright, *How the Laws of Physics Lie* (Oxford: Clarendon Press, 1983), 40–3.

tures as possible; and that is what the positivist will take hold of. I see *CC* as a tool for getting new causal information out of old. But for the positivist there is no right or wrong about the old causal information. *CC* functions rather as an effective, but not very perspicuous, way of summarizing some very complicated facts about what the total pattern of purely associational laws is like. In this case, any set of so-called 'causal laws' that produces the right constraints on the laws of association is as good as any other. Correlatively, one should be able to tell whether a set of causal laws is acceptable or not just by looking at what relationships the probabilistic laws bear to one another.

Chapter 3 shows that one has to look further. One needs to know some fundamentally causal information as well—one needs to know facts about singular causes. It is indeed possible to see causal laws as inference tickets from one probabilistic law to another—but only if the individual causal histories are right. There is thus a concealed *ceteris paribus* condition for the inference ticket to obtain; and that concealed condition brings in a notion of causality which is at the same modal level—or higher—than the one we are trying to eliminate. Exactly the same thing is true of capacities. So far we have not explored any of the caveats that need to be added to the schemes in the earlier sections of this chapter. When we do, we see that the constraints laid out there hold only *ceteris paribus*; and the conditions that must be added use concepts, like interaction, that are already deeply infiltrated by the concept of capacity itself.

To see how the need for caveats affects the program for modalizing away both causes and capacities in turn, consider another central problem in philosophy of science where a similar strategy has been tried—the problem of the existence of theoretical entities. Positivists, once they had given up wanting to translate theoretical terms, tried instead to construe theories as summaries of laws about observables. The recent work by van Fraassen[16] and Arthur Fine[17] on empirical adequacy supplied the kind of modal force necessary to make this program plausible, and to give to positivist theories the same kind of power and applicability as can reasonably be assumed for their realist counterparts. Realists maintain that a belief in the truth of the theory and the existence of the

[16] Op. cit.
[17] 'Unnatural Attitudes: Realist and Instrumentalist Attachments to Science', *Mind*, 95 (Apr. 1986), 149–79.

entities it employs is the best way to justify our expectations that the predictions of the theory will obtain, and also the best way to explain why they do. Van Fraassen and Fine recommend a belief that stops just short of that, namely a belief in the *complete empirical adequacy* of the theory, with emphasis on *complete*. Briefly, the belief in the *complete empirical adequacy* of a theory commits one to an open-ended set of counterfactuals that follows from the theory, though not to the truth of the theory itself. It is this open-ended commitment that accounts for novel predictions, justifies applying the theory in untested domains, and yields empirical laws beyond those already known. The theory is thus for Fine and van Fraassen to be treated as a valid inference ticket—a totally valid inference ticket—but its claims are not to be taken seriously as descriptions of the world.

This is the same kind of programme as that outlined in this chapter, and it faces the same kind of objection. The objection is nicely put by Wilfrid Sellars.[18] Theories are supposed to be summaries of laws about observables. But in fact there are no (or at least not enough) laws about observables to summarize. We do not need theoretical entities to explain the systematic behaviour of observables. Rather, they are necessary to systematize observable behaviour in the first place. Sellars's own example is of earthworm behaviouristics. There may well be certain regular things that earthworms do in response to different stimuli; and these will be accessible to a diligent behaviourist. But there is always a concealed *ceteris paribus* assumption: this is what earthworms do *so long as* they are neurologically sound. And neurological soundness is not something which can be behaviouristically defined. Sellars thinks the structure of this simplistic example is characteristic of the relationship between theory and observation. The point can be put quite generally. Theories do not have unlimited applicability; the domain and limitations on the domain can be constructed only by already using the theory and the concepts of the theory. Theories come before observables, or are at least born together with them. This forces us to give real content to the theoretical terms and not take them merely as part of a scheme, albeit a modal scheme, for summarizing laws about non-theoretical entities.

I do not here intend to start on a discussion of scientific realism in general, neither to defend nor to attack it. I introduce the subject

[18] *Science, Perception, and Reality* (London: Routledge & Kegan Paul, 1963), 1–40.

purely for formal reasons. For Sellars's defence of realism about theoretical entities is exactly parallel in structure to the argument I want to give for interpreting capacities realistically. Difficulties for the purely modal view of capacities arise from two main sources: first from *ceteris paribus* conditions involving interactions, and second from the need to control for multiple capacities associated with the same feature. I begin with interactions. Consider the argument in section 4.2, which carries one from causal laws true in one set of circumstances to those that must be true in others. That argument maintains that the first causal law gives evidence for a capacity, and that the capacity will exhibit itself in a new causal law in any new test situation. That assumes that the capacity remains intact. It is, of course, part of the point of taking capacities seriously as things in the world, and not just particularly strong modalities, that they should remain intact from one kind of situation to another. But that does not mean that there can be no exceptions; it means that any exception requires a reason. Probably the most common reason for a capacity to fail to obtain in the new situation is causal interaction. The property that carries the capacity interacts with some specific feature of the new situation, and the nature of the capacity is changed. It no longer has the power that it used to.

Causal interactions are a longstanding problem in philosophy of science. John Stuart Mill thought they were the principal reason why chemistry was far less successful than physics. When two forces in mechanics are present together, each retains its original capacity. They operate side by side, independently of one another. The resulting effect is a pure combination of the effect that each is trying to produce by itself. The law of vector addition gives precise content to this idea of pure combination. In chemistry, things are different. The acid and the base neutralize each other. Each destroys the chemical powers of the other, and the peculiar chemical effects of both are eliminated. This is not like the stationary particle, held in place by the tug of forces in opposite directions. When an acid and a base mix, their effects do not combine: neither can operate to produce any effects at all.

The generalizations, or inference tickets, of sections 4.2 and 4.3 need to be amended to allow for interactions. The principles should be prefixed, 'So long as there is no interfering causal interaction, then . . . '. But how is this to be done without admitting interactions into the descriptive content of the world? I think it cannot be done. I have argued already that, one modal layer down, the *ceteris paribus* conditions on the inference patterns licensed by causal laws cannot be specified without already invoking strong notions of causality. This same kind of problem arises here. Causal interactions are interactions of causal *capacities*, and they cannot be picked out unless capacities themselves can be recognized. The attempt to 'modalize away' the capacities requires some independent characterization of interactions; and there is no general non-circular account available to do the job.

Consider the usual statistical characterization of an interaction. For simplicity, I will discuss only the now familiar three-variable model. In this case the response variable z is supposed to be dependent on two independent variables x and y, plus an error term whose expectation is taken to be zero. In the conventional structures of Chapter 1, the dependence was assumed to be additive; but here we allow more general forms of dependency, and assume only that

$$z = f(x,y) + u$$

or

$$\text{Exp}(z) = f(x,y)$$

Then x and y have no interaction in their effect on z if and only if

$$f(x,y) - f(x',y) = g(x,x'), \text{ for all } y$$

and

$$f(x,y) - f(x,y') = h(y,y'), \text{ for all } x$$

That is, the difference or contrast in the mean response for two different levels x and x' of the first factor is independent of the level of the second, and conversely. Alternatively, instead of talking about the expectation of z, one could talk about the influence which x and y jointly contribute. In that case, to say that x and y do not interact is to say that the level of y does not affect the difference in contribution that comes from varying x, and the other way around as well.

This is the standard characterization of interaction.[19] Yet it is obviously of no help for the problem here. For we began with the commonsensical intuition that, at least in some special domains, one could determine the relationship between a cause and its effect with

---

[19] See S. R. Searle, *Linear Models* (New York: Wiley, 1971), s. 4.3(d).

all the other causal factors fixed at some set levels; and then one could make a meaningful prediction: so long as no interactions occur, the relationship between x and z will be the same at new levels of y as at the level already studied. The idea behind the commonsense intuition is that an interaction is a special type of thing, with an identity of its own, a kind of process which—like any other real process—can be identified in a variety of different ways, depending on the concrete situation. Taking the statistical characterization above to give the essence of what an interaction is trivializes this intuition; and in particular it trivializes the prediction one wanted to express, and deprives it of content. The claim that one can expect the relationship between x and z to be the same no matter what the level of y is, except in cases where x and y interact, now says no more than this: you can expect the relationship to be the same except where it is different; and this is no proper expression of the methodology actually in use. In practice one looks for independent evidence that an interaction is occurring, and some account of why it should occur between these variables and not others, or at these levels and not others. The chemistry examples are a good case. One does not just say the acid and the base interact because they behave differently together from the way they behave separately; rather, we understand already a good deal about how the separate capacities work and why they should interfere with each other in just the way they do.

Besides the characterization in terms of levels, interactions are also sometimes defined in the modelling literature in terms of the linearity of the structural equations. As remarked earlier, it is not really linearity in the variables that matters, but rather linearity in the parameters. That means that the influences should add: the factors x and y are additive in their influences if there is a function q (which is independent of y) and a function h (independent of x) and real numbers a, b, c such that

$$z = f(x,y) = a + bg(x) + ch(y)$$

But this definition does not help with the problem at hand either, for it is equivalent to the first: that is, it is a theorem that the factors x and y have no interaction if and only if they are additive.[20]

---

[20] For a proof of this theorem, see Mathias Klay, 'Interaction and Additivity in Experimental Designs', MS, Institute for Mathematical Studies in the Social Sciences, Stanford University, 26 May 1987. I would like to thank Mathias Klay for many helpful conversations, both about interaction and about regression methods in general.

What is needed for the positivist program to modalize away capacities is some separate, independent characterization of interaction, a characterization that employs no further problematic concepts like the concept of capacity itself. That is what we seem not to have, and for good reason, I have been arguing. For the concepts of capacity and interaction are genuine descriptive concepts, and are not in any way to be reduced to more primitive ideas. There are methods for determining when they obtain, but the methods cannot double as empirical reductions, for the two concepts are woven together in these methods and cannot be pried apart.

I describe only the interplay between these two concepts; but that is too simple. A large number of other equally non-reducible concepts are involved as well, concepts like those of enabling conditions, precipitating factors, triggers, inhibitors, preventatives, and the like. These are the kinds of concept that will have to go into a proper account of what capacities are and how they operate; that is, in some sequel to this book. I mention them just to make this point: given the rich fabric of all these interconnected concepts, we can make sense case by case of the methods we use to study any one of them and of the judgements we arrive at. But without them the whole enterprise of establishing and using any causal concepts at all will seem arbitrary, and even pernicious.

Rather than pursue these other concepts, the same point can be made by thinking again about an idea already discussed in a number of different places throughout this book—the concept of mixed capacities. Do you think the rain will result in more accidents or in fewer, this holiday weekend? The question is hard to answer, because the rain has different, opposing effects: by keeping people at home, it will tend to prevent accidents; but by worsening the conditions of the roads, it will tend to produce them. How, then, should one think about the question probabilistically? Principle CC gives a preliminary version of the connection between causes and probabilities: a cause should increase the probability of its effect, and a preventative should lessen it, when all other causes are held fixed. But Chapter 3 teaches a more complicated lesson. It is not sufficient to hold fixed just the other causal factors; one must rather hold fixed the operation of all the other *capacities* that may be at work as well, whether those capacities are attached to separate causal factors, or to the very one under investigation. Otherwise facts about capacities and facts about probabilities will have no systematic connection with each other. Again, the program to modalize away capacities founders. An ascription of a capacity cannot be taken merely as an inference ticket to get from one fact about probabilities to another, that is, as an efficient summary of complicated facts about the pattern of pure probabilities; for the pattern it summarizes is not a pattern involving just the probabilities themselves but a more variegated pattern, involving both probabilities and capacities in an essential way.

There is one sweeping reply to be made to this argument, the reply of what is often called 'the radical empiricist', and that is to reject the whole fabric of causal concepts, and all the layers of modality as well: there are just isolated empirical happenings, and nothing more. All the rest is talk. Applied to the case of theoretical entities and theoretical laws, this reply leads to the familiar doctrine of the radical underdetermination of theory by facts. The same is true for the fabric of causal concepts. Some interlocking sets of joint hypotheses about capacities, interactions, precipitating conditions, and the like will be ruled out; but always, it will be maintained, an infinity of equally satisfactory alternatives remains. In both cases, I think, the challenge to the radical empiricist is the same. What is so special about the elementary facts with which you are willing to begin? What characteristics do they have that singular causings or capacities lack? My own view is that once the foundational picture of knowledge and the infallibility of our sense experience have been renounced, there is no good answer to this question. It seems it must then be left to rest on some very implausible view of concept formation, and not to arise from any convincingly grounded strictures on how knowledge can be acquired or on what can count as justification. That is, radical empiricism is a doctrine either without ground, or grounded on mistaken premises. The most stringent kind of empiricism that seems to me to make sense is the empiricism of practice that I advocate throughout; the empiricism that demands that each quantity be measured and each claim be tested. And the principal argument of this book is that causes and capacities are as empirical in that sense as it is possible to be.

There is one further point about radical empiricism that I should like to make, and that is to stress what a poor reconstruction of science it provides. There is now fairly widespread agreement that Carnap's project[21] to build the claims of science systematically from

[21] Op. cit.

some acceptable empirical core upwards cannot work. Nothing remotely like the science that we have can be arrived at in this way. But there has recently been more optimism among empiricists for a strategy like van Fraassen's, which does not aim to reconstruct the claims of science from some data deemed acceptable as a starting-point, but rather takes the claims of science at face value, and aims only to fit the data into some single consistent model which exemplifies the laws and theories as well. Van Fraassen's book is after all called *The Image of Science*, and although he takes the name from a different source, this is commonly thought to be one of its chief advances: his semantic view of theories, together with his rejection of any empiricist doctrines of meaning, combine to give his programme the possibility of succeeding where Carnap's attempts at a logical construction of the world failed, in providing an image totally acceptable to an empiricist of much the same science that appears to everyone else.

That is the claim I quarrel with. It seems to me to contain far more illusion than fact. The most immediate thing to notice for the theses of this book is that there are no detailed examples that treat causal laws; so the programme of modalizing them away has not yet got any real attempts, let alone successes, to support it. Perhaps less apparent is the fact that the cases that have been studied, the cases that give content to the programme and make it seem plausible, are all essentially cases of fitting low-level theories into models of high-level theories, and not cases which handle the 'raw data' itself. That makes it easy to overlook two crucial problems. The first is the problem of how this 'raw data' is to be made to fit. There is perhaps the not fully articulated assumption that the data fit into low-level theory in exactly the same way that low-level theory fits into higher; and that the job of securing the fit must already have been carried out by the scientist in the process of justifying and accepting the theory. It remains to the philosopher just to reconstruct this work, to cast it into a form acceptable for a radical-empiricist image of science.

But that will not do, for even at the lowest level, science never treats of the kind of datum that the radical empiricist finds in the world. Scientific claims are tested, not against the empiricist's data (e.g. 'individual temperature readings from samples of lead, bubble chamber photographs') but rather against 'phenomena detected from them (e.g. the melting point of lead, the weak neutral current, changes in the rate of solar neutrino emission)'. The quotation is from a paper by James Bogen and James Woodward[22] which argues for just this claim. This is also one of the central theses of Ian Hacking's *Representing and Intervening*.[23] I introduce it here to make the point that scientific practice itself cannot be relied on to have already, in outline, produced a model that contains both the raw data and the hypotheses which, according to the radical empiricist, are supposed to cover the data. One can, of course, continue to insist that only the kind of datum admissible for a radical empiricist is relevant. But to do so would make even our most sensible and trustworthy tests look crazy and unmotivated. The tests suppose that nature is full, not only of data, but of phenomena as well, phenomena which—relevant to the point here—include capacities, interactions, and the like; and I do not see why we need to start with the assumption that this is false.

The second problem with focusing on the connection between lower- and higher-level theories is that it is easy in doing so to lose sight of the *ceteris paribus* conditions, some of which have played a central role in the arguments of this chapter. For the need for *ceteris paribus* conditions is felt most acutely—and by many felt only at all—when the theory is brought to bear, not on a model, but on a real, concrete thing. For models are often constructed to fit the laws exactly, with no need for *ceteris paribus* conditions.[24] Yet when one theory is related to another, it is usually by embedding the models of one in the models of another, and not by comparing the actual treatments the two would provide should they attempt to describe accurately the behaviour of concrete physical systems.

But here I do not want to dwell on these *ceteris paribus* conditions. Although I have devoted a lot of attention to them, I believe they are ultimately the wrong way to think about the problem. I used them as a kind of ladder to climb out of the modalization programme, a ladder to be kicked away at the end. They must be introduced if one is stuck with the project of reducing causings and capacities, or of modalizing them away. But I advocate giving up that programme

---

[22] 'Saving the Phenomena', *Philosophical Review*, xvcii (1988), 303–52.

[23] I. Hacking, *Representing and Intervening* (Cambridge: Cambridge University Press, 1983).

[24] N. Cartwright, 'The Simulacrum Account of Explanation', in *How the Laws of Physics Lie*. See also R. Giere, *Explaining Science* (Chicago: University of Chicago Press, 1988).

entirely—especially since it seems not to work—and accepting that capacities and causings are real things in nature. There is, I think, no other view of nature that can give an adequate image of science. In this doctrine I follow John Stuart Mill, whom I will discuss in the next section. I turn to Mill because my views are almost an exact replication of his, and discussing his arguments may help to make my own more clear.

4.5. Mill in Defence of Capacities

Keynes, in the discussion cited at the end of section 4.3, maintained that economic phenomena were probably not atomistic—that is, in the terminology of this book, economic life is not governed by stable capacities. John Stuart Mill believed that it was. I want to review here Mill's ideas about method and theory in economics to see how he arrived at this conclusion. Briefly put, Mill believed that the laws of political economy and the laws of mechanics alike are laws, not about what things do, but about what tendencies they have. This thesis should be familiar. Substituting the word 'capacity' for Mill's word 'tendency', his claim is exactly what I aim to establish in this book. Indeed, the observations that brought me to this conclusion, although set in a different context, are almost identical to the arguments that moved Mill. For the remainder of this section and through most of Chapter 5, I suggest that the reader take my 'capacity' and Mill's 'tendency' to be synonymous. No distinction will be drawn until section 5.6.

Mill's ideas about the structure and method of political economy are most explicitly described in his essay 'On the Definition of Political Economy and on the Method of Philosophical Investigation in that Science', written in the 1830s,[25] and in the chapter 'On the Logic of the Moral Sciences' in his *System of Logic*, which first appeared in 1843. Mill combined an original view on the structure of theories in political economy[26] with a well-known anti-inductivist view on method, which he shared with his father James Mill and other followers of Ricardo.

With respect to method it is important to note, as Samuel

[25] In *Collected Works*, iv (Toronto: Toronto University Press, 1967), 309–40.
[26] But note that what I call structure Mill called 'philosophical method'.

Hollander argues,[27] that for Mill the opposite of a pure inductive method was not a kind of 'a-priorism', but rather a 'mixed method of induction and ratiocination'. That means that the axioms from which scientific deductions begin are in no way arbitrary, but are rather to have two different kinds of support in experience: (1) there must be a 'direct induction as the basis of the whole', and (2) a kind of introspection or broad-based understanding of human nature, which is a source of information peculiar to the moral sciences. To illustrate:

Suppose, for example, that the question were, whether absolute kings were likely to employ the powers of governments for the welfare of or for the oppression of their subjects. The practicals would endeavour to determine this question by a direct induction from the conduct of particular despotic monarchs, as testified by history. The theorists would refer the question to be decided by the test not solely of our experience of kings, but of our experience of men. They would contend that an observation of the tendencies which nature has manifested in the variety of situations in which human beings have been placed, and especially observation of what passes in our own minds, warrants us in inferring that a human being in the situation of a despotic king will make a bad use of power; and that this conclusion would lose nothing of its certainty even if absolute kings had never existed or if history furnished us with no information of the manner in which they had conducted themselves.[28]

Mention of axioms and deduction may make it appear as if Mill endorsed the hypothetico-deductive method, which today stands opposed to inductivism, with the difference just noted that modern empiricists look only to the deductive consequences of a hypothesis for its empirical support, whereas Mill allows extra input for hypotheses in the moral sciences from our more generalized knowledge of human nature. But this is not the case at all. To see that, we need to look at Mill's views on structure, which complement exactly his views on induction and empirical support. Although Mill described his theories as deductive in order to dissociate them from the inductivist critics of Ricardo, who backed both a different philosophy of science and a different theory of political economy, it is not a structure that I would call deductive. For one cannot deduce even in principle what will occur in any future circumstance. With that Mill agrees:

[27] *The Economics of John Stuart Mill* (Oxford: Blackwell, 1985), ch. 2.
[28] Mill, op. cit., p. 325.

When in every single instance, a multitude, often an unknown multitude, of agencies, are clashing and combining, what security have we that in our computation *a priori* we have taken all these into our reckoning? How many must we not generally be ignorant of? Among those which we know, how probable that some have been overlooked; and even were all included, how vain the pretence of summing up the effects of many causes, unless we know the numerical law of each,—a condition in most cases not to be fulfilled; and even when it is fulfilled, to make the calculation trancends, in any but very simple cases, the utmost power of mathematical science with all its most modern improvements.[29]

Mill recognizes the pure deductive structure that goes along with our modern picture of the hypothetico-deductive method. He calls it the 'geometrical, or abstract method', and he rejects it. It seems he came to reject the geometrical method by reflecting on the weaknesses in his father's views, and in particular on T. B. Macaulay's criticisms of James Mill's 'Essay on Government'. He explains in his autobiography:

On examining . . . what the mind does when it applies the principles of the Composition of Forces, I found that it performs a simple act of addition. It adds the separate effect of the one force to the separate effect of the other, and puts down the sum of these separate effects as the joint effect. But is this a legitimate process? In dynamics, and in all the mathematical branches of physics, it is; but in some other cases, as in chemistry, it is not; and I then recollected that something not unlike this was pointed out as one of the distinctions between chemical and mechanical phenomena, in the introduction to that favourite of my boyhood, Thomson's *System of Chemistry*. This distinction at once made my mind clear as to what was perplexing me in respect to the philosophy of politics. I now saw, that a science is either deductive or experimental, according as in the province it deals with, the effects of causes when conjoined, are or are not the sums of the effects which the same causes produce when separate. It followed that politics must be a deductive science. It thus appeared, that both Macaulay and my father were wrong; the one in assimilating the method of philosophizing in politics to the purely experimental method of chemistry; while the other, though right in adopting a deductive method, had made a wrong selection of one, having taken as the type of deduction, not the appropriate process, that of the deductive branches of natural philosophy, but the inappropriate one of pure geometry, which, not being a science of causation at all, does not require or admit of any summing-up of effects. A foundation was thus laid in my

---

[29] Id., *A System of Logic* (1872), in *Collected Works*, vii. 460.

thoughts for the principal chapters of what I afterwards published on the Logic of the Moral Sciences; and my new position in respect to my old political creed, now became perfectly definite . . .[30]

This apparently led Mill to write the essay discussed above.

The difference between geometry and natural philosophy, or mechanics, can be put this way. The axioms of geometry are mutually consistent, whereas those of mechanics are not. A conclusion which is derived from some proper subset of the principles of geometry will remain true when further premisses are added, but not so with conclusions in mechanics. The laws of mechanics interfere with each other:

Among the differences between geometry . . . and those physical Sciences of Causation which have been rendered deductive, the following is one of the most conspicuous: That geometry affords no room for what so constantly occurs in mechanics and its applications, the case of conflicting forces, of causes which counteract or modify one another. In mechanics we continually find two or more moving forces producing, not motion, but rest; or motion in a different direction from that which would have been produced by either of the generating forces . . . what the one force does, the other, partly or altogether, undoes. There is no similar state of things in geometry. The result which follows from one geometrical principle has nothing that conflicts with the result which follows from another. What is proved true from one geometrical theorem, what would be true if no other geometrical principles existed, cannot be altered and made no longer true by reason of some other geometrical principle.[31]

Mill proposes, then, that political economy should be modelled on mechanics and not on geometry. But his view of mechanics is peculiar from the standpoint of a modern empiricist, who will construct mechanics itself on the model of geometry. This is implicit in Russell's remarks on causation, and it has been carefully and persuasively articulated in the standard American works of mid twentieth-century philosophy of science, notably by Ernest Nagel[32] and Carl Hempel.[33] To begin with, Mill takes the fundamental laws of nature (or more accurately, one major category of laws of nature, since he wants to allow different forms like conservation laws as

[30] *Autobiography*, in *Collected Works*, i. 169. Quoted in Hollander, op. cit., pp. 89–90.
[31] Id., *Logic*, pp. 887–8.
[32] *The Structure of Science* (New York: Harcourt, Brace & World, 1961).
[33] *Philosophy of Natural Science* (Englewood Cliffs, N J: Prentice-Hall, 1966).

well) to be laws about individual causes and their effects; and second, he thinks that individual causes in physics are atomic in the way Keynes described. Indeed, sometimes Mill even uses the same kind of language that I have used to describe the assumptions of econometrics: economics is lucky because it treats fixed capacities that are stable from one context to another.

Chemistry contrasts with physics in a second respect. Although the laws of chemistry are laws of cause and effect, the causes of chemistry are not governed by stable tendencies. Mill says:

A concurrence of two or more causes, not separately producing each its own effect, but interfering with or modifying the effects of one another, takes place, as has already been explained, in two different ways. In the one, which is exemplified by the joint operation of different forces in mechanics, the separate effects of all the causes continue to be produced, but are compounded with one another, and disappear in one total. In the other, illustrated by the case of chemical action, the separate effects cease entirely, and are succeeded by phenomena altogether different, and governed by different laws.[34]

Talk about interference and tendencies was in no way new with Mill. On the contrary, these were common ideas in political economy, where none of the laws proposed ever seemed to obtain in reality. Mill's contribution comes from his analysis of how interference works. In the first place, he stresses that one must not try to deal with exceptions by allowing laws that hold only for the most part:

we might have cautioned inquirers against too extensive *generalization*, and reminded them that there are *exceptions* to all rules . . . We have avoided the use of these expressions purposely because we deem them superficial and inaccurate.[35]

In another place he urges that it is a disservice to science to suggest that its laws have exceptions. The principal reason for this is that the fundamental laws (non-empirical laws, in Mill's terminology) do not hold for the most part, or even approximately for the most part; and conversely, those laws which are more or less true much of the time are not fundamental.[36]

[34] Mill, *Logic*, p. 440.
[35] Id., 'Definition', p. 337.
[36] This is a doctrine that I too have argued for: cf. Cartwright, *How the Laws of Physics Lie*.

The only laws that will be near to true for Mill are what he calls empirical laws. We are already familiar with the idea of an empirical law, as opposed to a fundamental one, from the discussion of Frisch and Haavelmo. The ideas translate fairly directly: what are empirical laws in Mill's terminology are laws which are low in autonomy for Frisch and Haavelmo. These laws hold because of some particular arrangement of background circumstances in a society through some period of time; and they remain true only so long as that particular arrangement persists. They thus have the limitation already noted when it comes to forecasting, for example, 'what would be the effect of imposing or of repealing corn laws, of abolishing monarchy or introducing suffrage, in the present condition of society and civilization in any European country'.[37] Mill says:

those immediate causes [in an empirical law] depend on remote causes; and the empirical law, obtained by this indirect mode of observation, can only be relied on as applicable to unobserved cases, so long as there is reason to think that no change has taken place in any of the remote causes on which the immediate causes depend. In making use, therefore, of even the best statistical generalizations for the purpose of inferring (though it be only conjecturally) that the same empirical laws will hold in any new case, it is necessary that we be well acquainted with the remoter causes, in order that we may avoid applying the empirical law to cases which differ in any of the circumstances on which the truth of the law ultimately depends.[38]

More important for my thesis, however, is not the fact that laws which are nearly true, albeit for particular situations and finite periods, are not fundamental, but rather that fundamental laws are not true, nor nearly true, nor true for the most part. That is because fundamental laws are laws about distinct 'atomic' causes and their separate effects; but when causes occur in nature they occur, not separately, but in combination. Moreover, the combinations are irregular and changing, and even a single omission will usually make a big difference. The philosophers who followed the geometric method, Mill continues,

would have applied, and did apply their principles with innumerable allowances. But it is not allowances that are wanted. There is little chance of making due amends in the superstructure of a theory for the want of sufficient breadth in its foundations. . . . That the deductions should be from

[37] Mill, *Logic*, p. 911.
[38] Ibid. 908.

the whole and not from a part only of the laws of nature that are concerned, would be desirable even if those that were omitted were so insignificant in comparison with the others, that they might, for most purposes and on most occasions, be left out of the account. But this is far indeed from being true in the social sciences. The phenomena of society do not depend, in essentials, on some one agency or law of human nature, with only inconsiderable modifications from others. The whole of the qualities of human nature influence those phenomena, and there is not one which influences them in a small degree. There is not one, the removal or any great alteration of which would not materially affect the whole aspect of society, and change more or less the sequences of social phenomena generally.[39]

This was one of the reasons why Mill was opposed to the statistical laws of Adolphe Quetelet's social physics. The statistics of social physics was modelled on astronomy, where the aim was to find the true orbit from a scatter of observations. The metaphysics was that of the major cause—either permanent or slowly evolving—like the arrangement of the other massive bodies in the heavens, and a myriad of minor, erratic, disturbing factors, which in a reasonable world would be independent and normally distributed around the true value so that their effects would average out. The Quetelet picture conceals a problem which Mill was deeply concerned about—a concern I share. Where in nature is the fundamental relation between a cause and its effect exhibited? In Quetelet's story of the major cause with its small, randomly distributed errors, it is easy to overlook the fact that the hypothesized relation never literally obtains for most causes. The cause is there, but some other different effect follows as a result of the interference of the minor causes. It is easier to overlook this problem from Quetelet's viewpoint than from Mill's: when the perturbing influences are small, at least the effect which occurs is close to the effect predicted, and although close is in no way good enough, it is somehow comforting. Mill could not indulge in this comfort. Society for Mill was nothing like the society of Quetelet. Causes did not divide themselves into major and minor, with the minor ones erratic and independent. All causes were subject to change, and the omission of even a single one from consideration could make a substantial difference.

This viewpoint of Mill is important to the question of capacities because it undermines a ready answer to my concerns. The question I pose is: what is stable in nature? Hume says that the association of

[39] Ibid. 893–4.

events is regular (or at least that, where science is possible, the association of events must be regular). But Mill notices that this is not so, neither in mechanics nor in political economy. What actually happens is highly variable, because of the shifting mixture of causes. That is why simple induction will not work as a method of discovery in these sciences. Specific combinations of causes do not stay fixed long enough to produce the data necessary for a good induction. Even should they do so, and should we be able to identify the particular mixture that obtains, the resulting law would not be the kind we seek. For it would tell us only what happens in that very specific arrangement. That information provides no good basis for decision, nor for prediction.

Nowadays there is a conventional answer to this kind of problem, also I think too ready and too glib; and that is the resort to counterfactuals: the requisite regularities may not in fact hold, but they are true counterfactually. In the language of the debate about Ricardo's abstract methods and strong cases, a language common again today, they hold '*ceteris paribus*'. That means they would hold *if* all disturbing causes were absent. But that will not do. That was Mill's own view, and one that I support as well. Even if these regularities did hold *ceteris paribus*—or, other things being *equal*—that would have no bearing on the far more common case where other things are *not* equal. Here are Mill's own words:

We might, indeed, guard our expression . . . by saying that the body moves in [the prescribed] manner unless prevented, or except in so far as prevented, by some counteracting cause, but the body does not only move in that manner unless counteracted, it *tends* to move in that manner even when counteracted.[40]

The solution that Mill offers is tendencies.

These facts are correctly indicated by the expression *tendency*. All laws of causation, in consequence of their liability to be counteracted, require to be stated in words affirmative of tendencies only, and not of actual results.[41]

This is a solution that Mill worries about. He is after all an empiricist, following in the tradition of those early British empiricists who in part defined themselves by their rejection of the mysterious and occult powers which they took to be typical of scholastic science.

[40] Ibid. 444.
[41] Ibid. 445.

Nevertheless, he is ready to distinguish *laws about tendencies*, i.e. fixed capacities, from the more common idea of *tendency laws*, i.e. laws that say what tends to happen, or happens for the most part. These latter would describe at least partial regularities, or perhaps, counterfactual regularities, and hence be true to the philosophy of Hume. But we have seen that this is not what Mill is about. For Mill,

With regard to exceptions in any tolerably advanced science, there is properly no such thing as an exception. What is thought to be an exception to a principle is always some other and distinct principle cutting into the former, some other force which impinges against the first force and deflects it from its direction. There are not a *law* and an *exception* to that law—the law acting in ninety-nine cases, and the exception in one. There are two laws, each possibly acting in the whole hundred cases and bringing about a common effect by their conjunct operation.[42]

This follows immediately after Mill's claim that the philosopher who makes a mistake by expecting the literal consequence of an observed cause to occur has generally not erred by generalizing too extensively, but by 'making the wrong *kind* of assertion, he predicted an actual result when he should only have predicted a *tendency* to that result—a power acting with a certain tendency in that direction'.[43]

How seriously must we take this idea of powers? I have been arguing that modern science takes them very seriously indeed: that our methods and our use of science presuppose that tendencies, or 'capacities', are real. I think the same is true in Mill's work as well; given Mill's other assumptions, his talk about powers must be taken literally. One quick, final way to see this is to contrast Mill's endorsement of tendencies in the passages already cited with another silly suggestion that he makes to solve the same problem.

Peter Geach complains of Mill that he is not constant to his own arguments for tendencies. Geach refers to 'this doctrine of tendencies, which we have found in Mill mixed up with an entirely incompatible Humian invariable-succession theory. . . .'[44] That is because, on Geach's reading, 'Mill retreats into saying that physical laws do not state what *does* happen, but what *would failing interference* happen; but this is to abandon the Humian position.'[45] In

[42] Mill, 'Definition', pp. 337-8.
[43] Ibid.
[44] P. Geach and G. E. M. Anscombe, *Three Philosophers* (Oxford: Blackwell, 1961), 103.
[45] Ibid. 102-3.

contrast to Geach, it is not the introduction of counterfactuals that I complain of, but a completely unconvincing use of factuals that Mill resorts to in the *Logic*. For sometimes Mill tries to avoid introducing tendencies by claiming that the laws of nature, read as exceptionless statements of pure regularities, are not false after all. Every time a particular cause occurs, so too does its effect, despite all appearances to the contrary. Each of the separate, atomic effects exists inside, or as part of, or somehow or other in connection with, the apparent resultant effect:

In this important class of cases of causation, one cause never, properly speaking, defeats or frustrates another; both have their full effect. If a body is propelled in two directions by two forces, one tending to drive it to the north, and the other to the east, it is caused to move in a given time exactly as far in *both* directions as the two forces would separately have carried it . . . [46]

This is an idea that seems to me to be silly. To claim that a motion exists in a body even when that body is at a standstill, and passes all the conventional empirical tests for being at a standstill, is to forsake empiricism, and to do so in a way that violates its fundamental tenets more seriously than the admission of powers or tendencies, for it severs the existence of the motion from all our standard methods of measuring motions. That is something that an empiricist should not allow. Given Mill's other views, that laws of nature are about causes and effects, that these laws allow the construction of forecasts, and so on, ultimately I think Mill's view has to be that the fundamental laws of nature are laws that assign stable tendencies to specific causes. Despite Hume and some of Mill's teachings in earlier sections of the *Logic*, laws are not uniformities or regularities in nature after all.

4.6. Conclusion

This chapter has argued for singular causal happenings and the capacities that make them possible, embedded in a rich fabric of other interrelated factors like interactions, or enabling and inhibiting conditions. Inevitably, because this book starts from a study of probabilistic causality, some elements of the fabric, like

[46] Mill, *Logic*, pp. 370-1.

interactions, have received a good deal of attention, and others, like enabling conditions, have received virtually none at all. Still, I have tried to argue, all are equally necessary, all equally irreducible, and all equally objectionable to the radical empiricist. I have also maintained that this last does not matter: radical empiricism, without the accompanying doctrines of infallibility and foundationalism, is no longer a position to be reckoned with. But I have not argued for this last claim; and to do so would take an excursion into epistemology far from the more detailed issues of probability and method with which I began. Rather I offer, in its stead, another version of empiricism that I think can be taken seriously—and has been by large numbers of scientists since the seventeenth century—the empiricism of testing and measuring, an empiricism already too demanding to admit much of modern theoretical science, especially physics, which is prone to be driven more by the needs of mathematics than it is by the phenomena. Nevertheless, it does not exclude tendencies and causes.

Still, to say this is to side-step a crucial criticism. For the testing of causal claims at any level—whether claims about a single happening, about a more generic causal law, or about capacities and their operations—necessarily presupposes some metaphysical assumptions that cannot be tested by the same stringent logic. Yet this in no way distinguishes these claims from any other claims about the world. This is obvious in the case of laws of association. No regularity of pattern will tell us that a law obtains unless we know enough to ensure that the connections involved are law-like to begin with. Otherwise we are forced back to the hypothetico-deductive method, and that method provides no test at all. I think the same is universally true. Even what are supposed to be the 'purest' empirical assertions, like 'this facing surface is red', employ concepts which cannot be given ostensively but only make sense relative to an entire structure of other concepts in which they are embedded. Nor can they be tested, as many hoped, by pure inspection, without a rich background of assumptions, both physical and metaphysical, assumptions not much different in kind from those necessary to test causal claims.

But these are familiar views, and ones which are not very controversial at the moment. What I have wanted to do in this chapter is to attack a position that seems far more widespread nowadays than that of radical empiricism, a view that tries to let in just enough metaphysics, but not too much—the view that adopts laws, so long as they are laws of pure association, but rejects causes and capacities. Usually the programme that goes with this view proposes to account for the systematic use of causal language in science by reconstructing it as an indirect way of describing facts about laws of association. The point of this chapter has been to argue that that project will not work. I want to conclude with my own account of why it will not work.

The chapter begins with a picture of layered modalities: first laws of association, then causal laws, then capacities. But section 4.4 argues that this picture is only half true. The capacities are more than modalities; they are something in the world. Where, then, does that leave the lower modal levels, and in particular the laws of association? The popular view, which I want to attack, takes these as nature's givens. It is by laying down laws that nature regulates the unfolding of events. But I want to urge a very different picture that is open to us once we admit capacities into our world. It is not the laws that are fundamental, but rather the capacities. Nature selects the capacities that different factors shall have and sets bounds on how they can interplay. Whatever associations occur in nature arise as a consequence of the actions of these more fundamental capacities. In a sense, there are no laws of association at all. They are epiphenomena.

This, it seems to me, will be the most powerful way to reconstruct a reasonable image of science; and it is this picture that I urge philosophers to take up and develop. Already there is one concrete exemplar available to show how it can work, in systems of simultaneous equations of the causal-modelling literature that have played a role throughout this book. There we see one way to represent capacities with precise quantitative strengths, and with a systematic pattern for their interplay: in general they are additive, with departures from additivity represented by interaction terms. What is important for the project of taking capacities as primary is that these models show how it is possible to begin with capacities and to end with probabilities which describe the associations in nature as a consequence.

One of the chief strengths of an ontology which takes capacities as fundamental and associations as secondary is the quite important one that it can hope to give a far more realistic picture than can the alternative. We all know that the regularity of nature so central to the more conventional picture is a pretence. It does not lie on nature's surface, but must be conceived, if there at all, to be somewhere in the deep structure. I, in general, advise against deep structures. John Stuart Mill gives an alternative account, and this account makes good sense of what we observe about regularities and the lack of regularities. Nature, as it usually occurs, is a changing mix of different causes, coming and going; a stable pattern of association can emerge only when the mix is pinned down over some period or in some place. Indeed, where is it that we really do see associations that have the kind of permanence that could entitle them to be called lawlike? The ancient examples are in the heavens, where the perturbing causes are rare or small in their influence; and the modern examples are in the physics laboratory, where, as described in Chapter 2, our control is so precise that we ourselves can regulate the mix of causes at work. Otherwise, it seems to me, these vaunted laws of association are still very-long-outstanding promissory notes: laws of association are in fact quite uncommon in nature, and should not be seen as fundamental to how it operates. They are only fundamental to us, for they are one of the principal tools that we can use to learn about nature's capacities; and, in fact, most of the regularities that do obtain are ones constructed by us for just that purpose.[47]

[47] Again, this theme of the construction of regularities is one of the leading ideas of Ian Hacking's *Representing and Intervening* (Cambridge: Cambridge University Press, 1983).

5

Abstract and Concrete

5.1. Introduction

For John Stuart Mill the basic laws of economics are laws about enduring 'tendencies' and not laws about what happens; that is, laws about capacities and not just about the sequence of events. This is because the laws treat causes singly, but reality rarely isolates them from one another in this tidy way. Mill also maintains that the laws of economics are 'abstract', and by that he means something different. For the use of 'abstract' points, not to the content of the laws, but rather to the methods by which they can be established: inevitably an element of the a priori must be involved. But his claims about method and about content are driven by the same problems. Recall section 4.5. How could one establish a law about capacities by pure induction? The relevant instances of the isolated capacities never occur. We cannot do experiments in political economy which guarantee that each cause operates separately in order to observe its natural consequences. Worse, the mix of causes is continually changing, and it is almost always impossible to ascertain which are present. In a situation like this, the conventional methods of induction will be insufficient. For Mill, they must be augmented by principles we can glean from our general knowledge of human nature. This 'mixed method of induction and ratiocination' is what he calls 'a priori'.[1]

The very same problems that force one to use a mixed method in 'going upwards' from experience to general principle reappear when one wants to turn about and 'argue *downwards* from that general principle to a variety of specific conclusions'.[2] These problems are the focus of this chapter; and most of the themes are already present in Mill. First he argues that the route downwards involves adding

---

[1] J.S. Mill, 'On the Definition of Political Economy', in *Collected Works*, iv (Toronto: Toronto University Press, 1967), 325.
[2] Ibid.

corrections to allow for the effects of the disturbing causes that may be at work in any given situation. Following an analysis by Leszek Nowak, section 5.4 calls this the process of *concretization*. It is important that the additions and corrections should not be arbitrary:

The disturbing causes are not to be handed over to be dealt with by mere conjecture . . . they may at first have been considered merely as a non-assignable deduction to be made by guess from the result given by the general principles of science; but in time many of them are brought within the pale of the abstract science itself. . . . The disturbing causes have their laws, as the causes which are thereby disturbed have theirs . . . [3]

The passage contains in brief the answer to an important Humean question: why do we need to introduce capacities to serve as the subjects of our laws? Is it not sufficient to recognize that these laws are not literal descriptions of what happens in actual circumstances, but rather some kind of abstraction? The answer is, in a sense, yes: it is sufficient to recognize that the laws are abstractions of a certain sort. But that fact does not do the work a Humean needs. For capacities cannot be so readily decoupled from this notion of abstraction, since the converse processes of abstraction and concretization have no content unless a rich ontology of competing capacities and disturbances is presupposed. This mirrors the conclusions of section 4.4 that capacities cannot be modalized away, and is also one of the main theses of this chapter.

A second thesis which I share with Mill concerns the source of our information about the disturbing causes for any given case. Some are given by theory, but many are not:

The disturbing causes are sometimes circumstances which operate upon human conduct through the same principle of human nature with which political economy is conversant. . . . Of disturbances of this description every political economist can produce many examples. In other instances, the disturbing cause is some other law of human nature. In the latter case, it can never fall within the province of political economy; it belongs to some other science; and here the mere political economist, he who has studied no science but political economy, if he attempt to apply his science to practice, will fail.[4]

The fact that a great many of the disturbing causes fall outside the domain of the science in question leads to serious problems when we

---

[3] Ibid. 330.
[4] Ibid. 330-1.

try to tie our theories to reality. For it means there is never any recipe for how to get from the abstract theory to any of the concrete systems it is supposed to treat. We have only the trivial advice, 'Add back all the causes that have been left out and calculate the total effect by combining the capacities.' Is this the best we can do? If so, it highlights how ineliminable capacities are from our image of science; but it makes for a rather impoverished philosophy. I call this 'the problem of material abstraction'. Although I will have much more to say about it, I do not really have a complete analysis of the problem. I aim here to demonstrate its importance.

That indeed is the principal point of this chapter. I am going to describe a variety of different kinds of abstraction in science, and especially in physics. My central thesis is that modern science works by abstraction; and my central worry is that philosophers have no good account of how. In many cases the abstractions can be taken as claims about capacities; and hence this thesis supports my doctrine of the reality of capacities. But it has wider implications than that. I have wanted to argue, not just for capacities, but against laws; and where abstraction reigns, I will suggest, laws—in the conventional empiricist sense—have no fundamental role to play in scientific theory. In particular, scientific explanation seems to proceed entirely without them. They are the end-point of explanation and not the source.

5.2. Idealization and the Need for Capacities

The problem which drives Mill—and me—to tendencies is sometimes identified by philosophers of science with the problem of Galilean idealization. I begin with a characterization by Ernan McMullin of the specific aspect of Galileo's methods which is most relevant:

The really troublesome impediments, Galileo said more than once, are the causal ones. The unordered world of Nature is a tangle of causal lines; there is no hope of a 'firm science' unless one can somehow simplify the tangle by eliminating, or otherwise neutralizing, the causal lines which impede, or complicate, the action of the factors one is trying to sort out. . . . it is this sort of idealization that is most distinctively 'Galilean' in origin.[5]

[5] E. McMullin, 'Galilean Idealization', *Studies in the History and Philosophy of Science*, 16 (1985), pp. 264–5 n. 3.

The conceptual project of constructing ideal models along Galilean lines runs in exact parallel with the empirical project of designing a decisive experiment. The logic of such experiments was discussed in Chapter 2. McMullin describes them thus:

Experiment involves the setting up of an environment designed to answer a particular question about physical processes. The experimenter determines how Nature is to be observed, what factors will be varied, which will be held constant, which will be eliminated and so on. This sort of manipulation is not always possible; experimental method cannot be directly applied in palaeontology or in astrophysics, for instance. The move from the complexity of Nature to the specially contrived order of the experiment is a form of idealization. The diversity of causes found in Nature is reduced and made manageable. The influence of impediments, i.e. causal factors which affect the process under study in ways not at present of interest, is eliminated or lessened sufficiently that it may be ignored. Or the effect of the impediment is calculated by a specially designed experiment and then allowed for in order to determine what the 'pure case' would look like.[6]

McMullin himself summarizes what the problem of Galilean idealization is:

In Galileo's dialogue, *The New Sciences*, Simplicio, the spokesman for the Aristotelian tradition, objects strongly to the techniques of idealization that underlie the proposed 'new science' of mechanics. He urges that they tend to falsify the *real* world which is not neat and regular, as the idealized laws would make it seem, but complicated and messy.[7]

This problem is often thought to have a variety of fairly obvious solutions. McMullin's own paper is an attempt to lay these solutions out in a clear and persuasive way. I want to explain in this section why these solutions, good for the purposes for which they are intended, are nevertheless not solutions to the right problem. For they are solutions to an epistemological problem about how we know what happens in ideal circumstances, and not to the question which bears on capacities, the question of why one can extrapolate beyond the ideal cases.

First, a word about terminology. 'Idealization' is a common word to use in discussing Galilean methods, and also in many discussions about the connection between models and applications in contemporary science. But my problem, and Mill's—and also, I think,

---

[6] Ibid. 265.
[7] Ibid. 247.

the principal problem in connecting models with reality—is not one of idealization, but rather one of abstraction; and that is indeed what the problem was called at the time of Mill and Ricardo. Briefly, I think we can distinguish two different processes of thought, the first of which I will call *idealization*, the second *abstraction*. What philosophers usually mean by 'idealization' nowadays is a blend of both. The conflation is not surprising, since the two are intimately linked. Usually, the point of constructing an ideal model is to establish an abstract law. But the law which is actually exemplified in the model is not the same as the abstract law which is thereby established. On the one hand, the abstract law seems to say more, since it is meant to cover a great variety of situations beyond those represented by the ideal model. On the other hand, as Mill pointed out, what it says seems to be literally true in not one of these situations, unless of course we take it to be a law about tendencies.

Here is how I want to distinguish idealization and abstraction for the purposes of this book: in idealization we start with a concrete object and we mentally rearrange some of its inconvenient features—some of its specific properties—before we try to write down a law for it. The paradigm is the frictionless plane. We start with a particular plane, or a whole class of planes. Since we are using these planes to study the inertial properties of matter, we ignore the small perturbations produced by friction. But in fact we cannot just delete factors. Instead we replace them by others which are easier to think about, or with which it is easier to calculate. The model may leave out some features altogether which do not matter to the motion, like the colour of the ball. But it must say something, albeit something idealizing, about all the factors which are relevant. In the end we arrive at a model of an 'ideal plane', ideal for purposes of studying inertial motion. Consider the ball rolling down the plane. To calculate its motion, you must know the forces in each of three orthogonal directions. The problem is undefined until all the forces are specified, so you cannot just omit them. You must set them to some value or other; otherwise you cannot study the actual motion.

By contrast, when we try to formulate Mill's laws of tendencies, we consider the causal factors out of context all together. It is not a matter of *changing* any particular features or properties, but rather of *subtracting*, not only the concrete circumstances but even the material in which the cause is embedded and all that follows from that. This means that the law we get by abstracting functions very differently from idealized laws. For example, it is typical in talking about idealizations to say (here I quote specifically from McMullin): the 'departure from truth' is often 'imperceptibly small', or 'if appreciably large' then often 'its effect on the associated model can be estimated and allowed for'.[8] But where relevant features have been genuinely subtracted, it makes no sense to talk about the departure of the remaining law from truth, about whether this departure is small or not, or about how to calculate it. These questions, which are so important when treating of idealizations, are nonsense when it comes to abstractions.

From the remarks of McMullin's that I just cited it will probably be clear why I want to make some distinction of this kind. When my problem of abstraction is assimilated to the problem of idealization, it is easy to think, erroneously, that one can solve the combined problem by developing some notion of approximate truth. But that does not work. That is why I stressed in the discussion of Mill the distinction between laws about tendencies or capacities, and tendency laws—laws that say what tends to happen, or laws that approximate to what happens. This distinction is particularly apparent in modern physics, where most of the effects we study now are very, very small. They make only a tiny contribution to the total behaviour of the system, so that the laws for these effects are very far from approximating to what really happens. Thus the laws in microphysics are results of extreme abstraction, not merely approximating idealizations, and therefore are best seen as laws about capacities and tendencies.

My thesis, then, is that idealization and abstraction are different, and that idealization would be useless if abstraction were not already possible. In the next section I will distinguish a couple of different senses of abstraction, one of which leads to what will be called 'symbolic representations'. The other leads to laws about tendencies. It is abstractions in the sense of tendency laws that concern me. My basic idea is that the method of Galilean idealization, which is at the heart of all modern physics, is a method that presupposes tendencies or capacities in nature. We can see at least what my thesis amounts to by considering typical discussions of idealization, and the questions they are designed to answer. First there is a familiar philosophical topic:

8 Ibid. 257.

(1) How can we know what would happen in ideal (i.e. not real) circumstances?

Most of McMullin's discussion addresses this question. A typical answer tries to extrapolate in some systematic way: you look at what happens as you get closer and closer to the ideal—for instance, you make the plane smoother and smoother—and then you try to take some kind of limit. What would happen if the plane were perfectly smooth? This raises a second question:

(2) Is the ideal some kind of limit of the real?

This question is the focus of the historian Amos Funkenstein, in a discussion I will describe below. My question is yet a third one:

(3) What difference does it make?

Imagine that what happens in some ideal circumstance is indeed a limiting case of what happens in a series of closer and closer approximations to it; assume, moreover, that we have some good idea of how to extrapolate the series to its limit. So now we know what would happen if the circumstances were ideal. What does that teach us? We think it teaches us a lot. But to think so is to believe in capacities.

I first came to see this point clearly in a historical setting by thinking about some claims that Amos Funkenstein makes about Galileo's work.[9] So that is the context in which I will try to explain it. Funkenstein thinks about what turns out to be my second question: is the ideal any kind of limit of the real? According to Funkenstein, herein lies the great advance of Galileo. In Aristotle and throughout the Middle Ages, ideal or counterfactual situations were thought to be incommensurable with reality. What happens in real cases, even as they get closer and closer to the ideal, has no bearing on what would happen were ideal circumstances to exist. I give the simplest of Funkenstein's examples, from Aristotle:

Aristotle shows that [motion in the void] is incommensurable, and hence incompatible, with any conceivable motion in the plenum. . . . Other things (force or weight) being equal, the velocity of a body moving (in analogy to forced motion) in the void must always be greater than the velocity of an equal body moving in a medium, however rare, since velocity increases in an

---

[9] A. Funkenstein, *Theology and the Scientific Imagination* (Princeton, NJ: Princeton University Press, 1986).

inverse proportion to resistance, that is, in direct proportion to the rarity of the medium. Nowhere does Aristotle suggest, as do many of his interpreters to this day, that because of this, motion in the void would be instantaneous or with infinite speed, only that it would be 'beyond any ratio.' The temptation is strong to render his intentions with the equation $\lim_{R \to O} F/R = \infty$ ($v = F/R$), but it would be wrong. He argues only that velocities in the plenum are commensurable in the proportion of their media, i.e., $v_1/v_2 = m_1/m_2$, and that this equation becomes meaningless when $m_2 = 0$ (void), since there is no proportion between zero and a finite magnitude. The movements of two equal bodies moved by equal forces in the void and in the plenum have no common measures.[10]

Here is how Funkenstein summarizes his argument:

Counterfactual states were imagined in the Middle Ages—sometimes even, we saw, as limiting cases. But they were never conceived as commensurable to any of the factual states from which they were extrapolated. No number or magnitude *could* be assigned to them. . . . For Galileo, the limiting case, even where it did not describe reality, was the constitutive element in its explanation. The inertial motion of a rolling body, the free fall of a body in a vacuum, and the path of a body projected had to be assigned a definite, normative value.[11]

I want to focus on the middle sentence: 'For Galileo the limiting case, even where it did not describe reality, was the constitutive element in its explanation.' I focus on this because it is a glaring *non sequitur*. Funkenstein has been at pains to show that Galileo took the real and the ideal to be commensurable. Indeed, the section is titled 'Galileo: Idealization as Limiting Cases'. So now, post-Galileo, we think that there is some truth about what happens in ideal cases; we think that that truth is commensurable with what happens in real cases; and we think that we thereby can find out what happens in ideal cases. But what has that to do with explanation? We must not be misled by the 'ideal'. Ideal circumstances are just some circumstances among a great variety of others, with the peculiarly inconvenient characteristic that it is hard for us to figure out what happens in them. Why do we think that what happens in those circumstances will explain what happens elsewhere?

I hope by now it becomes apparent: the logic that uses what happens in ideal circumstances to explain what happens in real ones is the logic of tendencies or capacities. What *is* an ideal situation for

10 Ibid. 158–9.
11 Ibid. 177.

studying a particular factor? It is a situation in which all other 'disturbing' factors are missing. And what is special about that? *When all other disturbances are absent, the factor manifests its power explicitly in its behaviour.* When nothing else is going on, you can see what tendencies a factor has by looking at what it does. This tells you something about what will happen in very different, mixed circumstances—but only if you assume that the factor has a fixed capacity that it carries with it from situation to situation.

The argument is structurally identical to my arguments in the last chapter that our standard ways of identifying, and then applying, causal models make sense only if one presupposes that there are capacities, stable capacities that remain the same even when they are removed from the context in which they are measured. I began with the econometricians because they are very explicit. But their methods are just a special case of the general method of Galilean idealization, which underlies all modern experimental enquiry. The fundamental idea of the Galilean method is to use what happens in special or ideal cases to explain very different kinds of thing that happen in very non-ideal cases. And John Stuart Mill has taught us that to reason in that way is to presuppose that there are stable tendencies, or capacities, at work in nature—as indeed is reflected in Galileo's own language. In *Two New Sciences*, for instance, he explains that he is trying to discover 'the downward tendency which a body has from its own heaviness'.[12] This description of the project is not a happy one for followers of Hume; but it is, I think, a correct one where 'Galilean' methods are employed.

5.3. Abstractions versus Symbolic Representations

There is, according to the last section, an intimate connection between idealization and abstraction, as these ideas were characterized there: one looks for what is true in an ideal model in order to establish an abstract law. But there is a difference between what is true in the model and the abstract law itself. For the ideal model does not separate the factors under study from reality, but rather sets them into a concrete situation. The situation may be counterfactual; still it is realistic in one sense: all the other relevant factors appear as

---

[12] G. Galileo, *Opere*, viii (Berkeley, Calif.: University of California Press, 1967), 258.

well, so that an actual effect can be calculated. What is ideal about the model is that these factors are assigned especially convenient values to make the calculation easy. Of course, the model is also to a great extent unrealistic. It is after all only a representation, and not the real thing, so necessarily much will be left out. In a scientific model the factors omitted altogether should be those deemed irrelevant to the effect under study. They do not need to be mentioned, because the values that they take will make no difference to the outcome.

But the difference between what is represented and what is not represented is not my main concern here. What I want to stress is the difference between the law which describes the happenings in the model and the abstract law which is thereby established. The first is a kind of *ceteris paribus* law: it tells what the factor does *if* circumstances are arranged in a particularly ideal way. In the abstract law the *ceteris paribus* conditions are dropped; and for good reason, since this law is meant to bear on what happens in the more frequent cases where conditions are not ideal. How are we to understand these abstract laws, laws that describe no real concrete situations nor any counterfactual ones either? I have been maintaining that, in certain special cases where the composition of causes is at stake, those laws are best thought of as ascriptions of capacity. Will that serve as a general solution? We must start by distinguishing between two different kinds of abstraction that occur in the construction of theories. The problem of how to reconnect the abstract with the concrete is on a quite different scale in the two cases. Only one admits the tidy solution of tendencies or capacities, and that is the one that matters here. The other sense can be found in Pierre Duhem's well-known discussion of abstraction. This section will describe Duhem's ideas in order to make clear the difference between the two senses.

Duhem used the claim that physics' laws are abstract as a weapon to attack scientific realism. But Duhem was no anti-realist in general. Rather, he wanted to draw a distinction between everyday concepts and the very special concepts of science, especially of mathematical physics. According to Duhem, the concepts of common facts are concrete; those of physics, abstract and symbolic:

When a physicist does an experiment two very distinct kinds of representations of the instruments on which he is working fill his mind: one is the image of the concrete instrument that he manipulates in reality; the other is a schematic model of the same instrument, constructed with the aid of symbols supplied by theories; and it is on this ideal and symbolic instrument that he does his reasoning, and it is to it that he applies the laws and formulas of physics.[13]

The two instruments can never collapse into one:

Between an abstract symbol and a concrete fact there may be a correspondence, but there cannot be complete parity; the abstract symbol cannot be the adequate representation of the concrete fact, and the concrete fact cannot be the exact representation of the abstract symbol.[14]

For Duhem, a single theoretical fact may correspond to a variety of concrete facts. For example, 'The current is on' corresponds to a collection of concrete facts like the displacement of a spot in a galvanometer, the bubbling in a gas volt-meter, the glow of an incandescent lamp inserted in the wire, and so forth. Conversely:

The same group of concrete facts may be made to correspond in general not with a single symbolic judgment but with an infinity of judgments different from one another and logically in contradiction with one another.[15]

In current philosophical jargon, the concrete facts underdetermine the theoretical. The simplest case has to do with the precision of measuring instruments: any of the incompatible values for a theoretical quantity that lie within the range 'of error' of the most precise instrument used correspond equally well to the facts available.

The abstractness of the objects that physicists reason about in contrast to the objects they manipulate comes from two sources. One is connected with issues of wholism, which philosophers nowadays commonly associate with Duhem. Theoretical concepts are abstract because they can only be applied to reality by presupposing an interconnected web of theoretical laws and rules of correction. The second source is connected with the methods modern physics uses to represent reality. Physics aims at simplicity of representation. But nature, as it comes, is complex and intricate. Hence there inevitably arises a mismatch between the abstract-theoretical representation and the concrete situations represented.

---

[13] P. Duhem, *The Aim and Structure of Physical Theory*, trans. P. P. Wiener (New York: Atheneum, 1962), 155–6.
[14] Ibid. 151.
[15] Ibid. 152.

The result is that the abstract formulae do not describe reality but imaginary constructions. They are best judged neither true nor false of reality itself.

An even more important problem than simplicity of representation, for Duhem, is the mathematical character of physics. The concepts of physics are precise, but reality itself does not exhibit this precise, quantitative nature:

Let us put ourselves in front of a real concrete gas to which we wish to apply Mariotte's (Boyle's) law; we shall not be dealing with a certain concrete temperature, but with some more or less warm gas; we shall not be facing a certain particular pressure embodying the general idea of pressure, but a certain pump on which a weight is brought to bear in a certain manner. No doubt, a certain temperature corresponds to this more or less warm gas, and a certain pressure corresponds to this effort exerted on the pump, but this correspondence is that of a sign to the thing signified and replaced by it, or of a reality to the symbol representing it.[16]

Notice here that Duhem is talking about the theoretical concepts of a purely macroscopic physics, and is not concerning himself with the status of tiny, unobservable entities. A second example that Duhem uses is the sun. In order to find a 'precise law of the motion of the sun seen from Paris':[17]

The real sun, despite the irregularities of its surface, the enormous protuberances it has, will be replaced by a geometrically perfect sphere, and it is the position of the center of this ideal sphere that these theories will try to determine. . . . [18]

This kind of falsification is for Duhem an unavoidable consequence of the mathematical formulation of physics. He seems here to be echoing the earlier criticisms of Laplace and Poisson directed against Fourier's elevation of differential equations to first place in physics. Differential equations require sharp geometric boundaries.[19] So long as the differential equations are taken, as by Poisson especially, merely as convenient summaries of results that can be rigorously established by summing the interactions of molecules, the reality of these boundary conditions is irrelevant. But the

---

[16] Ibid. 166.
[17] Ibid. 169.
[18] Ibid. 169.
[19] Norton Wise shows how to set Duhem in this historical context. To learn more about it, see his 'The Flow Analogy to Electricity and Magnetism', pt. 1, *Archives for History of the Exact Sciences*, 25 (1981), 19–70.

issue can no longer be ignored if we drop the molecular basis and adopt the differential equations as primary. When we do so, we inevitably end up with a physics whose concepts no longer describe reality, but are irreducibly 'abstract and symbolic'.

There is a tendency nowadays to see Duhem's problem as one of approximation.[20] This, indeed, as McMullin points out, was Salviati's response to Simplicio: the geometrical configurations of the new world system may not be perfectly realized in nature; still, geometry provides the right way to describe nature. One must just be careful to allow for the impediments of matter when applying the theory to real cases: 'The errors lie then, not in the abstractness or concreteness, not in geometry or physics as such, but in a calculator who does not know how to keep proper accounts.'[21] McMullin acknowledges how controversial this view was. Mathematics itself was on trial as the language of physics. But in the end McMullin endorses the view that the problems are ones of approximation, which can be calculated away:

Mathematical idealization has worked well for the natural sciences. The extent to which it *is* an idealization has steadily diminished as the mathematical language itself has become progressively more adapted to the purposes of these sciences. It would be hazardous today to argue . . . that there are causal factors at work in the natural world that are *inherently* incapable of being grasped in mathematico-physical terms. The weight of the inductive argument is surely in the opposite direction. But it should be underlined once again that what has made this possible is not so much the reducibility of the physical as the almost unlimited plasticity of the mathematical.[22]

Not all modern mathematical physicists have been as optimistic as McMullin. I have already mentioned Laplace and Poisson, who saw special problems in the mathematics of differential equations. Maxwell and Kelvin had even stronger views. They both despaired that any mathematics accessible to the human mind could represent nature even approximately as it is. I describe the barest outlines of their views to make clear the philosophical point. The reading I give of both Maxwell and Kelvin comes from Norton Wise, as does my reading of the debate between Laplace and Poisson, and Fourier; the

---

[20] Besides McMullin, cited in section 5.2, see also 'How the Laws of Physics Don't Even Fib', *PSA [proceedings of the biannual Philosophy of Science Association meetings] 1986* and also the study of approximation by R. Laymon, *The Scientific Use of Idealizations and Approximations*, forthcoming.

[21] McMullin, op. cit., p. 251.

[22] Ibid. 254.

full story can be found in Crosbie Smith and Norton Wise's biography of Kelvin.[23] The views of Kelvin and Maxwell, according to Wise, were rooted in the problem of free will, of how mind can act on nature without violating any law of nature. They saw a possible solution in a continuum theory of the physical world. Yet even an infinitesimal portion of a true continuum would require an infinite number of variables for a complete mechanical specification, which therefore lay outside human capacity. Furthermore, no portion of a continuum could be regarded as isolated from the remainder, so that only a physics that considered the behaviour of objects in physical connection to the infinity of their surroundings could be considered realistic. Finally, motion in the continuum would present essential instabilities and essential singular points, both incalculable and therefore beyond the scope of mechanical determination and of human knowledge.

In order to obtain a workable description of thermodynamic phenomena in this situation, Maxwell compounded idealizations to produce a statistical mechanics. A dedicated mechanical philosopher, Maxwell nevertheless thought statistics and probability theory a necessary addition. In *How the Laws of Physics Lie*,[24] I described how the falsehoods inevitably introduced by the constraints of any given representational scheme can sometimes be mitigated by the introduction of still more falsehood. For example, imagine that one wants to stage a play about the signing of the Declaration of Independence and to make it as realistic as possible; that is, to try to reproduce the events of the signing, as far as one can, on the stage. Literal reproduction will be impossible because of the demands of the theatre. Two participants whisper, conspiring together over some strategy. But the whispering actors cannot be heard, so they stand up and leave the main table, to walk to the side of the stage and speak aloud to each other. I think this is just how Maxwell saw statistical mechanics. Although, considered in itself, the introduction of probabilistic laws was a significant misrepresentation of nature, it was nevertheless the best device to couple to the equally misrepresenting atomistic, or non-continuous, mechanics, in order to save the phenomena.

For Kelvin, too, matters came to a head over questions of thermo-

[23] *Energy and Empire: A Biographical Study of William Thomson, Lord Kelvin* (Cambridge: Cambridge University Press, 1988), chs. 11, 12, and 18.
[24] (Oxford: Clarendon Press, 1983).

dynamics and reversibility. He wanted to give up the ordinary *abstract dynamics* of points and forces where, as Wise explains, 'total reversibility was conceivable in the idealized world of an isolated, finite system of discrete particles and conservative forces', and replace it by a *physical dynamics* more suited to the infinite complexity of the real world, a dynamics where 'reversibility necessarily failed in the infinite real world of imperfectly isolated systems of friction and diffusion phenomena'.[25] But even physical dynamics would still get the essential description of nature wrong, for, as Kelvin said, 'The real phenomena of life infinitely transcend human science'.[26]

Both Duhem and Kelvin used the word 'abstract' to describe a mathematical physics that they believed could not adequately picture the world, even by approximation. Duhem also uses the word 'symbolic'; and it is this label that I shall adopt. For I should like to reserve the word 'abstraction' to pick out a more Aristotelian notion, where 'abstraction' means 'taking away' or 'subtraction'. For Aristotle we begin with a concrete particular complete with all its properties. We then strip away—in our imagination—all that is irrelevant to the concerns of the moment to focus on some single property or set of properties, 'as if they were separate'. The important difference is that Aristotle's abstract properties are to be found in the real world whereas, in the view of Kelvin and Duhem, the symbols of physics do not correspond to anything in nature. For Aristotle, at least in principle, all the properties that science will study are there in the objects to begin with.

I say 'in principle' because this claim too is itself some kind of idealization. Consider a triangle, a real triangle drawn on a blackboard. Even when the chalk and the colour and all the other incidental features are subtracted, the shape that is left is never a real triangle. But let us pretend that it is, and use the label *abstraction* for the process of isolating an individual characteristic or set of characteristics in thought, for that will allow us to set aside one extremely difficult problem while we focus on another. The central question I want to pursue is: what do abstract or symbolic claims say about reality? However this question is to be interpreted, surely the answer must be easier for claims which are abstract in the narrower

[25] C. Smith, and N. Wise, op. cit., ch. 18.
[26] Quoted in ibid.

Aristotelian sense than for those which are symbolic. Since I assume that capacities are as much in the world as anything else, symbolic claims are not so immediately relevant to my project here. So I propose in the remainder of this chapter to concentrate just on the abstract in Aristotle's sense.

Even then the focus needs to be narrowed. For one can in the Aristotelian way abstract all sorts of different features and think about them in all sorts of different ways for different purposes. This is apparent in Aristotle, for the same property can be considered differently by different sciences. The triangular may be considered *qua* geometrical shape; for instance: are all triangles trilateral? But one may ask different questions and abstract for different purposes. Galileo, for example, speculated on its role in the substructure of matter: are the primitive parts of fire triangular in shape? If the triangles were sharply pointed, they could penetrate into other matter and thus account for why things expand on heating. For the purposes here I want to focus on the kinds of abstraction that end in laws of capacity, where the factor to be isolated is a causal factor and the question to be posed is, 'What does this factor produce by its own nature?' It should be clear by now that I have replaced the essentialism of Aristotle with its talk of natures by the more humdrum concept of capacity, and thereby translated the question into a new one: 'What capacities does this factor possess?' Much of this book has been dedicated to showing how we can answer that question, and in a way that should satisfy any reasonable empiricist. The problem of this chapter, and especially the next section, is almost converse to that one: Given a proposed answer, how is it to be understood? What does the abstract statement of capacity say about any concrete things?

5.4. What do Abstract Laws Say?

What do abstract laws say about real things? There are two aspects to this question, which may or may not be the same depending on one's philosophical perspective:

(1) What facts in the world make abstract claims true?
(2) How do abstract laws bear on the more concrete and more descriptive laws that fall under them?

To attack the first question directly is to become involved in ancient problems about the relations among universals or among abstract entities. Does the property of being a gravitational force have the second-level property of carrying a capacity? Or is this rather a way of categorizing the property of force? And in either case, what is the relation between the property itself and the fact which is supposed to be true of it? I want provisionally to set aside these traditional metaphysical issues and to focus instead on the second question, which bears more immediately on the day-to-day conduct of science.

I want to consider the relationship between the abstract capacity claim and the vast network of individual facts and concrete laws that fall under it. By concrete laws I mean ones which fill in, in their antecedent, all the relevant details; so that they can be read as literally true or false in the most straightforward sense. For short, I call this subordinate network of concrete laws the 'phenomenal content' of the abstract capacity law. The question I want to pose is: 'What determines the phenomenal content of an abstract law, and how does the phenomenal content bear on the truth of the law?'

The discussion has several parts. The first, 5.4.1, is a brief survey of some of the literature in semantics that treats of similar problems. But I think no help is to be found there. Essentially, the linguistics literature either reverts to a kind of simplified Platonism or wanders in the same quandaries and with the same kinds of question that I raise here. I include two specimens of recent discussions among philosophers of language to illustrate. The second section 5.4.2, describes the work of Leszek Nowak on abstraction which I think offers the best formulation available of how abstract and concrete laws relate. The last section, 5.4.3, raises a new problem that is not apparent from Nowak's scheme, one which opens new and difficult philosophical questions.

5.4.1. Semantics for Generics and Habituals

There is a new literature in semantics that might be of help, for abstract-tendency laws seem to work in much the same way as ordinary generic sentences, like 'Dogs bark', or habituals, like 'John smokes', and to create many of the same problems. In all three forms it is equally difficult to say what connection there is between individual cases and the more general truth they fall under. It is natural to suppose some kind of quantificational account, but it seems this is no more suitable for more general generic forms than it is for capacity laws. Should the quantifier be 'all' or 'most', or something weaker? None of these works for all of the following:

Dogs are mammals.
Dogs eat meat.
Dogs give milk to their young.

The list is taken from Greg Carlson's 'Generic Terms and Generic Sentences'.[27] One might try to solve the problem it presents by assuming that the meaning of generic sentences is to be given through a disjunction of different quantifiers. But, as Carlson argues, this strategy will not work, since 'the predicted ambiguity fails to materialize'. A conventional solution for capacity laws is to affix some restriction about normal conditions. But the solution fails, for, as sections 4.5 and 5.2 remark, the conditions in which capacities reveal themselves in their canonical behaviour are usually in no sense normal at all; it requires the highly artificial and contrived environment of a laboratory to manifest them. Nor does 'normalcy' seem any better for generics in general, for consider

Frenchmen eat horsemeat.[28]

A typical strategy for linguists trying to construct a semantics for generics and habituals is to introduce abstract entities. Carlson is a good example.[29] *Dogs* in 'Dogs bark' is a kind-denoting term; and the John of 'John smokes' is some kind of abstract construction from the more primitive space–time instances of John which may get involved in specific episodes of smoking. This raises a technical problem. Once there are two different types of John there must equally be different types of smoking, since the enduring individual John is not a proper subject for smoking in the episodic sense, any more than the kind *dogs* can bark in the same sense that an individual dog can. The trick, then, is to see that predicates too come in two types: gnomics and episodics; and that this distinction is grounded in a difference in reality:

One plausible answer is that our (conceptual) world consists of two very

---

[27] *Journal of Philosophical Logic*, 2 (1982), 148.
[28] Ibid. 156.
[29] See also A. Kratzer, 'An Investigation of the Lumps of Thought', MS, University of Massachusetts at Amherst.

different sorts of things. First, there is the phenomenal world of colours, sounds, motion, and so forth—the world of space and time. But there is also the *organization* of that world; one of its fundamental organizing principles is that of individuation. The episodic sentences are about the space–time world; gnomic sentences are about the organization of that space–time world, the individuals viewed as an integral part of the organization.[30]

The semantic problem becomes trivial then. 'John smokes' is true if and only if the abstract entity *John* has indeed the structural or organizational property designated by the gnomic 'smokes'.

Carlson's is a particularly accessible and straightforward account. But this simple idea, employing Platonic entities and the familiar correspondence theory of truth, is at the heart of a great number of the recently proposed treatments in linguistics. Whether reasonable or not, it is of no help to our problem here. Nor do linguists like Carlson expect it to be. He likens it to another case. For Carlson, 'is red' is a dispositional predicate. He argues

What, then, of our semantic representation of a sentence claiming x is red? There seems little point in going beyond a simple formula like $Red(x)$. This makes the claim that x is in the set of red things. How we go about figuring out if x is red or not is our problem; not the semantics. We cannot ask that the semantics have built into it a theory of epistemology.[31]

Alice ter Meulen is a philosopher-linguist with a different attitude. Her model-theoretical analysis of generics is developed in the framework of situation semantics. For ter Meulen, the kind of information that Carlson consigns to epistemology is instead at the core of the semantics. In situation semantics, the truth of a sentence depends on the structure of situations relevant to the sentence, and these are picked out by the context. On ter Meulen's account, the connection between the generic truth on the one hand and the specific episodes that it relates to on the other is reflected in the selection of situations to include in the semantic evaluation: 'Although generic information is not directly descriptive of particular situations, its purpose is to classify such situations as being of a particular type and as entering into relations with other types and situations.'[32] As a crude example, given the information

---

[30] Carlson, op. cit., p. 168.

[31] Ibid. 160.

[32] A. ter Meulen, 'Generic Information, Conditional Contexts and Constraints', in E. C. Traugott, A. ter Meulen, J. S. Reilly, and C. A. Ferguson (eds.), *On Conditionals* (Cambridge: Cambridge University Press, 1986), 125.

that John smokes, the situation of having John house-sit for a week is one that you—being someone who dislikes the lingering smell of cigarettes—will want to avoid. To turn to the case of capacity laws, the information that massive bodies fall should lead me to expect that I will have to use a magnet or a piece of tape to keep Emmie's paintings affixed to the refrigerator door. According to ter Meulen, understanding the generic consists primarily in 'getting attuned to' the constraints that classify situations in the right way. Again, this does not help to solve our problem, but only to relabel it: 'What kinds of constraint are associated with capacity laws, and where do they come from?'

5.4.2. Abstraction and Concretization

It is probably well, then, to take Carlson's advice and to look back to the philosophical literature for a solution to what is not really a semantic but rather a philosophical problem. The most detailed work on abstraction has been done by Leszek Nowak, as part of his reconstruction of Marx's thought in *Das Kapital*. In this work Nowak gives an account, a very straightforward and simple one, of how to get from an abstract law to the concrete laws that make up its phenomenal content. Although the account does not solve all the problems, it seems to me to be right in its basic outline. But no Humean will be able to admit this. For Nowak's story involves the obvious idea that one must add *corrections* and *additions* as one moves from the abstract to the concrete. It is critical to the account that these corrections should not be *ad hoc* addenda just to get the final results to come out right: they must be appropriately motivated. I take it that means they must genuinely describe other causes, interferences, impediments, and the like. But it follows from that that the scheme can only work if we are already in control of a rich set of non-Humean, capacity-related concepts. This provides an important argument in favour of capacities. For we cannot get by without an account of how abstraction works in science, and an account like Nowak's is the most natural one to give. Or, more strongly, I would say in its basic outline it is the correct one to give; and it is an account that presupposes an ontology of capacities.

Let us turn to a description of what Nowak does. Throughout his work Nowak tends to use the terms 'idealization' and 'abstraction' interchangeably. Indeed, his book is titled *The Structure of Ideali-*

zation: Towards a Systematic Interpretation of the Marxian Idea of Science,[33] and it early on presents the problem in the conventional way in terms of *ceterus paribus* conditions and counterfactuals: 'Such an economic system', says Nowak of the system described by Marx, 'resembles ideal gases, perfectly rigid bodies and other constructs of the type. In other words, the law of value is satisfied vacuously in the empirical domain investigated by Marx. It is therefore a counterfactual statement.' Moreover, claims Nowak, Marx himself clearly intended his work that way: 'He built that law according to his own principle of abstraction',[34] and he cites Marx:

The physicist either observes physical phenomena where they occur in their most typical form and most free from disturbing influence, or, wherever possible, he makes experiments under conditions that assure the occurrence of the phenomenon in its normality.

But

In the analysis of economic forms . . . neither microscopes nor chemical reagents are of use. The force of abstraction must replace both.[35]

Despite his use of both terms, Nowak is undoubtedly, in his work on Marx, primarily discussing the problem which I have called the problem of abstraction, and not the problem of idealization. For his concerns are structural, and metaphysical: what is the structure of Marx's theory? What should the laws—like the law of value—be taken to say? They are not immediately methodological, nor epistemological: how can one decide whether the law of value is true or not? As we shall see, one of the principal objections to Nowak's views turns on the fact that his scheme does not give any clues about how to approach the epistemological questions.

Nowak's central idea is the one I have been endorsing here. The basic scientific laws do not literally describe the behaviour of real material systems. They are instead to be taken as abstract claims. What that means is to be explained in part by a proper account of scientific explanation: not hypothetico-deductive, as the positivists would have it, but rather on account with the structure of a 'method of abstraction and idealization'. Nowak proposes that the fundamental laws of science give the essential behaviour of a factor, or, in

---

[33] (Dordrecht: Reidel, 1980).
[34] Ibid. 9.
[35] Quoted in ibid. 9.

Mill's terminology, describe its tendencies. The essential behaviour of a factor will be revealed in certain ideal circumstances. So the method begins with an *idealizational statement*, i.e. one of the form

T^k:

if $G(x)$ and $p_1(x) = 0$ and ... and $p_k(x)$
$= 0$, then $F(x) = f_k(H_1(x), ..., H_n(x))$

where $G(x)$ is some realistic description, $p_i(x) = 0$ is an idealizing assumption meant to describe counterfactual conditions, and the function f gives the essential connection between the factor F on the one hand and the factors $H, H_1, ..., H_n$ on the other. Then the idea is simple. Step by step, the idealizing assumptions are to be replaced by conditions that obtain in nature, until finally a true description is obtained which describes the actual relationship between F and H_1, ..., H_n in real concrete objects. The next lower stage looks like this:

T^{k-1}:

if $G(x)$ and $p_1(x) = 0$ and ... and p_{k-1}
$= 0$ and $p_k(x) \neq 0$ then $F(x)$
$= f_{k-1}((H_1(x), ..., H_n(x), p_k(x))$
$= g[f_k(H_1(x), ..., H_n(x)), h(p_k(x))]$

It is important to note that the new relationship, g, between F and the Hs is a functional of the previous one, f_k, plus the function h, which 'expresses the impact of the factor p_k on the investigated magnitude F (*correctional function*)'.[36] At last 'in the limiting case of *final concretization* all the idealizing assumptions are removed and all the appropriate corrections are introduced':

T^0:

if $G(x)$ and $p_1(x) \neq 0$ and ... and $p_k(x) \neq 0$, then $F(x)$
$= f_0(H_1(x), ..., H_n(x), p_k(x), ..., p_1(x))$
$= n[f_1(H_1(x), ..., H_n(x), p_k(x), ..., p_2(x)), m(p_1(x))]$[37]

Although Nowak in the first instance describes the method of abstraction and idealization as the method followed by Marx in *Das Kapital*, his central thesis is far stronger: that this is a method in use throughout the sciences, in fact probably the principal method. I agree. Section 4.3 has already argued that a structure like this is at the foundation of our conventional use of causal reasoning in the social sciences. It is typical in physics as well. I want here to develop a

[36] Ibid. 29   [37] Ibid.

point that I have made before.[38] My previous discussion concerned quantum mechanics, and I continue to use that case to avoid having to repeat the details. But quantum mechanics is intended just as an example; the structure I describe depends in no way on the peculiarly quantum features of quantum mechanics. It is equally characteristic of classical mechanics as well, as Ronald Giere points out in his recent book.[39]

The behaviour of the quantum state ϕ is governed by the Schroedinger equation:

$$H\phi = -i\hbar\partial\phi/\partial t$$

H is called the Hamiltonian. It is supposed to describe the possible energy configurations of the system, which can be thought of as the modern-day equivalent of Newton's forces. When we want to set quantum mechanics to describe the actual behaviour of some concrete system, where will the H come from? Essentially it is built up by pieces. At the start in quantum mechanics one learns a series of model Hamiltonians: e.g. the Hamiltonian for the hydrogen atom, for the square-well potential, for the harmonic oscillator, and so forth. Later, more and more sophisticated systems are included. These Hamiltonians are not literally true to their names: they are certainly not the Hamiltonians for any real hydrogen atoms, or real harmonic oscillators. Instead, they tell what contribution is made to the total Hamiltonian by the fact that the system is a simple proton–electron pair, or a certain kind of simple spring. That is, they report what energy configurations hydrogen atoms, harmonic oscillators, and so on *tend* to have, that is, what their capacities are. If one wants to talk in terms of what energy potentials they *do* reveal in a concrete situation, then something like the following is necessary: the Hamiltonian (H) for any real hydrogen atom will be the sum of a principal function (call it H_0) for the abstract model of a hydrogen atom and then a separate, 'correction' term (H') which is motivated by the particular circumstances. This is usually written $H = H_0 + H'$.

Calling the last term 'a correction' is an important reservation. For of course any Hamiltonian function H can be written as a sum of

[38] Cartwright, op. cit.
[39] Ronald Giere, *Explaining Science* (Chicago: University of Chicago Press, 1988).

some given Hamiltonian plus a remainder ($H = H_0 + [H - H_0]$). What is necessary for the claim 'H_0 is the Hamiltonian for the hydrogen atom' to have the intended sense is that the remainder, H', should be the appropriate Hamiltonian to describe the capacities of the remaining relevant factors. Given the theory, it is not too hard to say what one means by that: the remainder should also come from one of these long lists of Hamiltonians prescribed by the theory. This of course conceals the essential and interesting question of what justifies including a particular Hamiltonian on the list. The answer involves studying the detailed methodology of capacity ascriptions, case by case, which is a natural sequel to the project of this book.

The discussion has now clearly returned to one of my recurring themes, that the account of what capacities are cannot be reductive, but instead requires the whole paraphernalia of other capacities, preventatives, correctives, and so forth. This fact has been a major source of criticism of Nowak's accounts of abstractions. For it is quite apparent that, without such robust attendant concepts, anything can explain anything by the method of abstraction and concretization.[40] But it is not really a fair criticism of the scheme taken as a general picture of scientific explanation, for it is clear in Nowak's account that the corrective functions are meant to be just that, *corrective*, and not arbitrary. The matter stands differently when one displays the scheme to defend the basic laws of Marx's *Das Kapital*. For there the work is of a different nature. It is to be hoped that in the end a more detailed general account of the nature of corrective factors *vis-à-vis* the principal ones can be given. But failing that, in each case study it needs to be made clear that in this or that concrete situation the designated factors are indeed correctives or preventatives, as required for the reconstruction, and also why that is true. I do not want to pursue the question of whether Nowak's scheme succeeds in doing this for *Das Kapital*, but rather turn to a much more general problem with the process of concretization that affects all the sciences.

5.4.3. Material Abstraction

The abstract law is one which subtracts all but the features of interest. To get back to the concrete laws that constitute its pheno-menal content, the omitted factors must be added in again. But

---

[40] This was pointed out to me in conversation by Stephan Armsterdamski, who attributed it to a colleague of his.

where do these omitted factors come from? I have already described the answer I believe in: given a theory, the factors come *from a list*. But the list provided by any given theory, or even by all of our theories put together, will never go far enough. There will always be further factors to consider which are peculiar to the individual case. I call this the 'problem of *material abstraction*'. I will outline the problem in this section and give several examples of the complexities involved.

On Nowak's account it is Marx's theory that tells what kinds of factor have been eliminated in arriving at the law of value. The same is true for quantum mechanics. The Hamiltonian for a 'real' hydrogen atom is supposed to be arrived at by adding correction terms to the ideal Hamiltonian, where the correction terms come from the theory itself, from the list of other acceptable Hamiltonians. This is not to say that the list of Hamiltonians is closed. On the contrary, much of the day-to-day work in quantum physics goes into enlarging the list. But the point is that, in principle at least, no correction can be added to fit an ideal Hamiltonian to a 'real' case unless we are prepared to include the correcting Hamiltonian on the list and to keep it there.[41] I have put 'real' in quotes to signal my worry. For I think that, no matter how open-ended the list is, this kind of process will never result in an even approximately correct description of any concrete thing. For the end-point of theory-licensed concretization is always a law true just in a model.

I can perhaps explain the point best by considering an example. We have already had much discussion of lasers, so I will use them as an illustration. I hope this will also serve to reinforce the claim that the problems connected with abstract capacity laws are not peculiar to the social sciences, but trouble physics as well. Recall section 2.2. The fundamental principle by which a laser operates is this: an inversion in a population of atoms causes amplification in an applied signal; the amplification is narrow-band and coherent.[42] This is a good example of a capacity law in physics. It is a law that subtracts all the possible interferences and tells what an inversion does in the abstract; that is, it tells about what capacity the inversion has. Inversion means there are more atoms in the upper state than in the lower, contrary to the way atoms come naturally. This already

---

[41] When we are not prepared to do this, the correcting factors are called 'pheno-menological', and the theoretical treatment is judged not to be complete yet.

[42] Cf. A. E. Siegman, *Lasers*, (Mill Valley, Calif.: University Science Books, 1986), s. 1.1, 'What is a Laser?'.

reminds us that accounts of capacity laws in terms of 'normal conditions' have the wrong starting idea, since the conditions must be very abnormal before the causal antecedent is ever instantiated.

The example serves a number of purposes. First, it concerns a capacity which we learned about at a particular time, since it is new with quantum mechanics. In classical mechanics, inversions could not produce signal amplifications. Second, the language appropriate to the manifestation of capacities—that of harnessing them—fits perfectly here. Quantum physicists have known since Einstein's 1917 paper on the A and B coefficients that inversions *could* amplify. But not until the microwave research of the Second World War did anyone have any idea of how to produce an inversion and harness its capacity.

Third comes a point that bears on my central concern with the process of concretization: how and why are features added back in? We do know ways to produce and use inversions now. So suppose you want to amplify some signal. It won't do just to ring up the shop and order an inverted population. What if it is delivered in a brown paper bag? Now you might be more cautious. You may know some general defeating conditions for amplification. For instance, the inverted atoms must not have a strong absorption line to another upper state from the lower state of the inversion. In fact, physicists take this term 'absorption', which has a particular concrete meaning here, and use it as a catch-all, to describe anything that might take up the energy that the inversion should yield—scattering, thermal agitations, or whatever. So imagine even that you use this catch-all phrase, and order an inverted population (at a given frequency), adding the feature that it must be in a non-absorptive medium (for that frequency). Won't you want to say more?

The answer is yes, but the added features will depend upon the vagaries of the concrete situation. What you will want to say depends intimately on what medium the inversion is in and how it was brought about, as well as on specific characteristics of the signal you want to amplify. For example, depending on what the inverted medium is, the temperature of the amplifier may have to be very finely tuned to the temperature of the source. One of the early reports by P. P. Kisliuk and W. S. Boyle[43] on successful amplification in a ruby laser provides a good illustration. Kisliuk and Boyle

[43] P. P. Kisliuk and W. S. Boyle, 'The Pulsed Ruby Maser as a Light Amplifier', in J. Weber (ed.), *Lasers* (New York: Gordon and Breach, 1968), 84–8.

report a gain factor of 2, using a second ruby laser as a source for the signal. The use of a laser source gave them a signal strong enough not to get lost in the noise generated by the amplifier. To get significant amplification, the frequency of the signal should be matched to the transition frequency in the amplifying ruby. But the transition frequency in ruby depends on its temperature, so the temperature of the source and the amplifier must be about the same. Not exactly, however, for the ruby line width is also temperature-dependent, and the gain is a function of the line width; so the optimum condition for a given temperature of this source is reached, not when the amplifier is at exactly that temperature, but rather when it is somewhat colder. The authors report, 'The expected difference is of the order of a few degrees near room temperature. . . . As the temperature is lowered, its control becomes more critical because of the narrowing of the line.'[44] This gives a sense of the kind of context dependence that affects the list of features which must be added back during concretization.

There is nothing special about the laser case. Consider a second example. Donald Glaser built the first bubble chambers, using diethyl ether. He operated on the principle that a passing charged particle has the capacity to cause bubbling in a liquid in a super-heated state. (The liquid is ready to boil and just needs a stimulant.) He was also successful with hydrogen and most of the heavy liquid hydrocarbons, like propane or freon. But, surprisingly, the bubble chamber did not work with xenon. Here the passing charged particles excite optical transitions, and the energy is sent off as light rather than producing the heat necessary to trigger the boiling.[45] Again, mundane facts about actual materials and their construction made a difference, facts outside the domain of the initial theory about the behaviour of superheated fluids.

After a while, it seems, in any process of concretization, theor-etical corrections run out and the process must be carried on case by case. How are we to describe this problem? A more precise for-mulation would be easy if we could be convinced that the boundaries between the sciences reflect real differences in their domains and are not arbitrary consequences of our interests, talents, and social

---

[44] Ibid. n. 3, p. 87.
[45] Cf. P. Galison, 'Bubble Chambers and the Experimental Workplace', in P. Achinstein, and O. Hannaway, *Observation, Experiment and Hypothesis in Modern Physical Science* (Cambridge, Mass.: MIT Press, 1985), 309–74.

organization. Distinct domains would at least allow us to circumscribe sets of features for given concretizations. This indeed is part
of what Mill was up to. Why, after all, was he writing on the *definition* of political economy, and why do questions about tendency
laws play a major role in that discussion? In part because they allow
him to give an account of what makes a theory in political economy
adequate. Ricardo's abstract principles (or improved versions of
them) could be counted true if, taken together, they provide a
'correct' account of the contribution that the factors of political
economy make to what happens in the closed domain of the
economy. But that does not mean that anything that can be derived
from Ricardo's theory need ever be literally true, for causes from
other domains can—indeed, always do—make a difference.[46]

Much the same is true in modern science. Limited domains,
however achieved, could circumscribe the features that need to be
added during the process of concretization. Our econometrics
models serve as an example. Could the equations ever be true, even
in the view of the most optimistic econometrician? Here the error
terms are a salvation. For they can be taken to represent, not only
random fluctuations or factors just overlooked, but also omitted
factors of a very special kind—those outside the domain of
economics. The weather is the usual example. So in a sense econometrics has a trick that gives its laws a chance of being literally true.
It has one factor—the unmeasurable error term—that picks up
anything that may have an influence, even though what it represents
may vary in an entirely unsystematic way from case to case.

Among other modern sciences, it is mechanics alone that can
claim to be able to write laws that are literally true. For it assumes
that 'force' is a descriptive term, and adopts the pretence that all
changes in motion are due to the operation of forces. A conventional
example illustrates: billiard balls are supposed to exemplify laws
about the conservation of momentum. But if someone shakes the
table, or a child grabs a ball and runs away, the predicted motion will
not follow. No matter, says the advocate of mechanics, the actions
of the child or of the quavering table can be redescribed in terms of
forces, and then the actual path of the ball will no longer seem to
violate the laws of mechanics.

---

[46] This is admittedly a rather one-sided account of Mill's project, designed to help
make clear my philosophical point.

The cosy realm of mechanics may be an exception. But everywhere else there is no avoiding the problem of how to complete the process of concretization once theory gives out. I have labelled this 'the problem of *material abstraction*', with the laser example in mind. A physicist may preach the principles by which a laser should operate; but only the engineers know how to extend, correct, modify, or side-step those principles to suit the different materials they may weld together to produce an operating laser. The physics principles somehow abstract from all the different material manifestations of the laser, to provide a general description that is in some sense common to all, though not literally true of any.

But perhaps this focus on physics and how it attaches to the world is misleading. Mill saw the same kind of problem in political economy. There capacities are real, but abstract in just the sense I am trying to characterize: no amount of theory will ever allow us to complete the process of concretization. Each individual case sets its own criteria of relevance. This is the problem I just called 'the problem of material abstraction'. Yet, in political economy, talk about the material system, despite its familiarity, seems just a metaphor. Despite these reservations, I propose to keep to the idea that it is somehow the material that is subtracted in the first step up into theory, since this opens the possibility of using the traditional scheme of Aristotle's four causes to tie together a variety of different kinds of abstraction that occur in science. These are the topics of the next section.

Before that I should issue a warning. I realize that I have not here given a systematic or complete account of the problem that concerns me. What I have done, I hope, is to point to it clearly so that others can make out its importance and pursue it as well. My pointing has not been at all neutral. I chose deliberately the Aristotelian language of matter, form, and function because these terms are fundamental to a preliminary description of the phenomena that appear in my image of science. This language is a threat to the neo-Humean covering-law theorist, and it is meant as such. I think we cannot understand what theory says about the world nor how it bears on it until we have come to understand the various kinds of abstraction that occur in science, as well as the converse processes of concretization that tie these to real material objects. For the neo-Humean, the explanatory structure of nature consists in a set of natural laws which regulate the sequence of events. But I look for an explanatory structure that will make sense of the methods of abstraction and idealization which are at the heart of modern science, and mere laws are not enough for this difficult job. What kind of different explanatory structure is required? That is a question to pursue once the topics of this book are concluded. It is at least clear that a philosophy of science that uses in an essential way the language of 'underlying principles', 'capacities', 'causal structure', 'forms', and 'functions' is no philosophy of science for a neo-Humean world.

5.5. Concreteness and Causal Structure

Material abstraction, the subject of the last section, is crucial for the successful construction of a scientific theory, and it raises interesting and troubling questions about truth and realism in physics. But it is by no means the only kind of abstraction that occurs there. I want in this last section to discuss other kinds of abstraction that take place after the subtraction of matter, to see how they are to be understood. The discussion will centre around some examples that have long perplexed me. A laser is a more concrete object than a damping reservoir; a crystal diode rectifier is likewise more concrete than a reservoir. Why? Most obviously the laser is an object rich in properties, whereas the damping reservoir is practically bare; the crystal diode rectifier is far more fully specified than the reservoir. Certainly sheer number of features makes some difference, I shall argue, but it is not primarily what matters. A far more significant difference depends on the kinds of feature specified, and not on the number. Damping reservoirs (which I will describe in more detail below) are objects identified entirely in terms of their output: they are supposed to eliminate correlations in systems they interact with. But nothing is specified about how they produce this result. Similarly, a rectifier is supposed to turn alternating currents into direct ones. The crystal diode rectifier does the same job, but in a highly specific way. We not only know more about the crystal diode rectifier than about the reservoir; what we know is how it works. We know its causal structure. This, I shall argue, is why the crystal diode rectifier is significantly more concrete than the reservoir, and the laser than the reservoir.

The arguments and ideas of this section are taken from a paper written—and conceived—jointly by Henry Mendell and me.[47] Since there is no possibility of attributing the pieces separably to either of us, I shall follow the opposite strategy and reproduce the final joint product as closely as possible.

Scattered throughout the works of Aristotle we find expressions such as 'things in abstraction' or 'things from abstraction',[48] where, as described earlier in this chapter, 'abstraction' means 'taking away' or 'subtraction'. According to Aristotle, many sciences study objects which have properties subtracted. We begin with a particular substance and all its properties, both accidental and essential. For instance, if we examine a triangle in the context of a geometric discussion, we may begin with a triangular marking on a slate. We then subtract the stone and chalk, the colour, and other properties incidental to being a triangle. What is left when this psychological process is complete is the triangle which is to be treated as if it were substance. The subtracted properties may be conceived as incidental to what the object is treated as, namely a triangle. Of course, what the object actually is, a stone slab, will be among the accidents of the triangle. In other words, abstraction amounts to a psychological isolation of certain aspects of the slate.

Unfortunately, a direct application of Aristotle's concept to lasers and reservoirs would seem to involve us in awkward questions of counting. For Aristotle, the less abstract object is closer to substance than the more abstract. How are lasers closer to substance than reservoirs? By virtue of having fewer inessential properties stripped away? If we adopted this view, we would be faced with the unpleasant task of figuring out how to count properties. A simple example illustrates the difficulties. Suppose we examine two markings on the blackboard, one triangular, the other quadrilateral. We examine the first as a right triangle, the second as a rhombus. To examine the right triangle we must subtract lengths of the sides, and the two acute angles, except such properties as are necessary for the markings to be a right triangle. To examine the rhombus, we subtract lengths of the sides, except the relative lengths of opposite

[47] N. Cartwright and H. Mendell, 'What Makes Physics' Objects Abstract?', in J.T. Cushing, C.F. Delaney, and G.M. Gutting (eds.), *Science and Reality* (Notre Dame, Ind.: University of Notre Dame Press, 1984), 134–52.

[48] See M.D. Phillippe, 'Aphairesis, prosthesis, chorismos dans la philosophie d'Aristote,' *Revue Thomiste*, 4 (1948), 461–79.

sides, and angles, except the relative size of opposite angles. Are rhombi more or less abstract than right triangles? The question, as would any answer to it, seems pedantic and baroque. For reasons such as this, it is not possible to develop a coherent total ordering for Aristotelian abstract objects. Some entities, such as rhombi and right triangles, are not ordered with respect to each other.

But there are natural ways of partially ordering the abstract. In his discussions of different sciences whose objects are known in abstraction, Aristotle suggests a simple criterion for ordering the relatively abstract and concrete. For example, the objects of geometry and mathematical astronomy are both known in abstraction. But the latter are nearer to substance, and hence are more concrete. Why? It is not because, if we count the properties subtracted in each case, we see that one set is smaller than the other. Rather, since geometry is used in mathematical astronomy, in doing mathematical astronomy we do not subtract geometric properties. But in doing geometry we subtract some properties which are in mathematical astronomy. For example, mathematical astronomy treats of the movement of stars. A geometric treatment of the same objects would not consider the movement, though it might treat the stars as fixed points, and the paths of the stars as circular lines. Hence, Aristotle claims, mathematical astronomy is nearer to substance. For it works with a more complete description of an object, and so comes closer to a description of what the object is.

Even within a science it is possible to see that certain objects are more abstract than others. Triangles are more abstract than right triangles, since the right angle of the right triangle is subtracted when it is treated just as a triangle. Yet all the properties of triangles apply to right triangles. We have, therefore, a means for partially ordering entities: A is a more abstract object than B if the essential properties, those in the description of A, are a proper subset of the essential properties of B.

In a similar way Aristotle orders sciences. Mathematical astronomy studies geometric properties of certain moving points. We can distinguish those properties which pertain to movement and those which pertain to geometric spaces. These constitute two different classes which are basic to human understanding. Thereby we can claim that the objects of science A are more abstract than those of science B if the set of kinds of property treated by A is a proper subset of the kinds of property treated by B. The basic insight is that, where there are two clear classes of properties, the entity with properties from one class will be more abstract than the entity with properties from both; that is, the entities can be ordered by looking at how their properties are nested.

The idea that we can partially order the abstract with reference to the nesting of kinds of property is intuitively reasonable, since the entity described by more kinds of property, in this very straightforward way, does seem to be more completely described. We cannot, however, literally apply these two Aristotelian orderings to the case of reservoirs and lasers: their properties are not nested, nor are they items in different sciences in a hierarchical ordering. Nevertheless, the Aristotelian notion gives us hints. Later in this section, this basic intuition will be deployed to develop a new notion of abstractness which fits the contrast in our paradigmatic case between lasers and reservoirs. But first it is necessary to isolate what the difference is that I have been trying to illustrate with my examples of lasers and reservoirs.

Section 5.4 described one obvious way in which the objects that physicists reason about may be abstract. They may lack stuff, and be made of no material. Contrast helium atoms, infra-red radiation, and electrons, on the one hand, with levers, oscillators, and reservoirs, on the other. The first are characterized in large part by what kinds of stuff they are made from; two electrons, two protons, and two neutrons; or electromagnetic waves between .75 and 1000 microns in length; or negatively charged matter. But its stuff is no part of what makes an oscillator an oscillator or a lever a lever. These objects are characterized in terms of what they do, and they may be made of any material that allows them to do what they are supposed to.

Reservoirs are another good example of this second kind. Reservoirs play a role in a variety of physical theories, especially thermodynamics and electromagnetics. Fundamentally, reservoirs are large objects, in one way or another, that swamp the systems they interact with. With respect to electricity, the earth is a reservoir that 'grounds' moving charge; in thermodynamics, a heat reservoir will give up heat until it finally brings the system it interacts with to its own temperature.

For specificity, the discussion here will concentrate on reservoirs in quantum optics, where the reservoir serves to erase the memory of systems with which it interacts. Correlatively, the reservoir is given

only enough formal characteristics to justify this outcome mathematically. To erase the memory in a system, one wants circumstances that would justify introducing a Markov approximation
there, and interaction with the reservoir licenses just this. The
reservoir is assumed to have a very great number of degrees of
freedom. Because it is large and chaotic, correlations among its own
states persist for only a very short time, and it in turn destroys
correlations in the systems to which it couples. Nothing else is
necessary for an object to be a reservoir.

The reservoir is a paradigm 'black box'. We know only its final
effects, but nothing of what goes on inside. Intuitively, it is a very
abstract object. But what sense of abstractness expresses this
intuition? Reservoirs undoubtedly lack stuff. Sometimes the electromagnetic field may serve as a reservoir (for oscillating atoms, for
instance); sometimes the walls of a cavity; sometimes the environment in a room. But a large number of objects that physicists reason
about lack stuff, and yet most are far more concrete than the
reservoir. Consider a laser. Lasers are supposed to amplify light and
produce a highly coherent beam of radiation. They too, like the
reservoir, can come in a variety of substances. They may be made of
rare earth ions floating in dissolved molecules, or from organic dye
solutions, or from ruby rods, or numerous other materials. Yet
certainly there is a sense in which the laser is a far more concrete
object than a reservoir. It is that sense of *concrete* that I am trying to
explain.

What is it that makes a laser a laser? The answer to this question
will provide the key to my problem about the abstract and the
concrete. The first operating laser was a ruby laser made in 1960 by
T. H. Maimon. Any introductory discussion will point out that ruby
is not essential. For instances:

Since 1960 many different kinds of lasers have been made. They do not all
contain ruby, nor do they all use a flashtube [as Maimon's did]. But what
they do have in common is an *active material* (e.g. the ruby) to convert some
of the energy into laser light; a *pumping source* (e.g., the flashtube) to
provide energy; and mirrors, one of which is semi-transparent, to make the
beam traverse the active material many times and so become greatly
amplified.[49]

[49] L. Allen and D. G. C. Jones, *Principles of Gas Lasers* (New York: Plenum Press,
1967), 2.

We see that a laser consists of at least three distinct components: the active material, the pump, and the mirrors. A more theoretical treatment will also add the damping reservoir as another essential component in this list (see Fig. 5.2, p. 000). In one sense, then, the specific make-up of the laser is irrelevant: no particular material is required. But in another sense the make-up is essential. The components are a part of the characterization of a laser as a laser.

But the specification of the components is not sufficient. Lasers work in a very special way. Recall from section 2.2. that in normal circumstances far more atoms in a given population will be in the ground state than in the excited state. The trick in making a laser is to create a 'population inversion' in which the situation is reversed. Quantum theory then teaches that the radiation from an initial atom de-exciting will stimulate others to radiate as well, and the radiation that is produced in this way will be highly coherent, unlike light from ordinary sources. This method of production is even referred to in the name: 'laser' stands for '*l*ight *a*mplification by *s*timulated *e*mission of *r*adiation'.

It is, then, part of the concept of the laser that it consists of a specific set of components which produce highly coherent light by a specific causal process. The method of operation for a reservoir, by contrast, is completely unspecified. The relevant degrees of freedom that are represented mathematically are not identified with any concrete quantities, and no account is given of the process by which these quantities couple to the smaller system and destroy its correlations. This, we think, is why a reservoir is intuitively so much more abstract than a laser. Reservoirs are specified only in terms of their outputs, whereas lasers are specified, not just in terms of their output, but also in terms of the constituent parts and the causal processes by which the output is produced.

It should be noted that this variety of abstractness is independent of, in a sense orthogonal to, the concept of the abstract as symbolic attributed to Duhem in section 5.3. For Duhem, concepts in physics are abstract because they are not faithful to the objects they represent. But, as with any descriptions in physics, constitution and causal structure may be treated with more, or less, faithfulness. The real ruby laser that Maimon built had a particular concrete causal structure. But so too did the larger environment in which it ran that served as a reservoir to erase correlations in it. How realistically these structures can be copied in the physicist's model depends on the fit between reality and the mathematical needs of the theory; and even intricately detailed causal structures can be extremely symbolic in Duhem's sense. Nevertheless, in a different sense, even a highly symbolic object is more concrete with a causal structure than without one.

The conclusion reached so far about our specific example is this. A laser is characterized by its structure and its internal causal processes, as well as its output, unlike a reservoir, which is characterized only by its ouput. Mendell and I follow Ernan McMullin, to whom our joint paper is dedicated, in thinking this is a crucial difference. For causality is at the core of scientific explanation. That is obviously not a surprising thesis to find here, after the four earlier chapters of this book. For McMullin[50] the key to explanation is not subsumption, or derivability from law, as it is in the conventional deductive-nomological account. Laws, in his view, may play a role in helping to trace the causal processes, but what explains phenomena in typical theoretical explanations are the structures and processes that bring them about. As can be seen at the end of Chapter 4, my view is even stronger. The laws themselves are generally pieces of science fiction, and where they do exist they are usually the objects of human constructions, objects to be explained, and not ones to serve as the source of explanation. Causes and their capacities are not to play a role alongside laws in scientific explanation, but—at least in many domains—to replace them altogether.

Whether one wants to take the stronger view that I advocate, or the weaker view of McMullin—weaker, but still strong enough to offend the Humean—the crucial point is to establish the connection between causation and explanation, for that is essential to the story Mendell and I want to tell about abstraction. For we start from the assumption that to add to a scientific model a description of an object's structure and causal mechanisms is not just to add some information or other, but to add crucial explanatory information. In fact it is to add just the information that usually constitutes explanation in modern theoretical sciences.

There are two different intuitive ways that one object may be more concrete than another, which Mendell and I combine in our account.

[50] E. McMullin, 'Two Ideals of Explanation in Natural Science', in H. Wettstein (ed.), *Midwest Studies in Philosophy* (Minneapolis, Minn.: University of Minnesota Press, forthcoming); and E. McMullin, 'Structural Explanation', *American Philosophical Quarterly*, 15 (1978), 139–47.

The first is the simple idea that an object is more concrete if it has more properties. Of course we do not know how to count properties. But we can still exploit this basic intuition by using the Aristotelian idea of nesting. If the properties used to describe one object are nested in those of another, the first description is less complete and, in that sense, more abstract than the other.

The second idea is that a description of an object is more concrete when we know, not just more about it, but more about what it really is under that description, that is, what is essential in that way of describing or treating the object. Again, we must face the problem of what counts as knowing more about what an object really is, and again we can take a solution from Aristotle. For Aristotle, the first and central role of a science is to provide explanatory accounts of the objects it studies and the facts about them. In each science the nature of the explanations may be different. In physics in particular, the basic objects, material substances, are to be explained in terms of the four causes. Julius Moravcsik[51] suggests that we regard the four causes of Aristotle as different factors in an account explaining what a substance or a system really is: source (efficient cause), constituent (matter), structure (form), and perfection or function (final cause).

The fullest explanation will give as many factors as may be appropriate. An account which specifies all four factors will give all the relevant information for identifying the material substance and for understanding what it truly is. For example, if the account is to be of a particular person, the fullest account will mention all four factors: parentage, flesh and bones, human soul, and what an active human soul does, in terms both of the growth of the person into a complete human and of the performance of a human soul. Other information is not relevant to understanding what this substance is as a person. Less information will be defective. Not all sciences, however, will use all explanatory factors. If we are to explain an Aristotelian unmoved mover, the constituent will be left out, because the unmoved mover is not encumbered with material constituents. Mathematics, on the other hand, is concerned with neither source nor function.

The four causes of Aristotle constitute a natural division of the sorts of property that figure in explanation, and they thus provide a way of measuring the amount of explanatory information given. We

[51] 'Aitia as Generative Factor in Aristotle's Philosophy', *Dialogue*, 14 (1975), 622-38.

can combine this with the idea of nesting to formulate a general criterion of abstractness: (*a*) an object with explanatory factors specified is more concrete than one without explanatory features; and (*b*) if the kinds of explanatory factor specified for one object are a subset of the kinds specified for a second, the second is more concrete than the first. Notice that we are here nesting *kinds* of factor and not the individual factors themselves, and that, as with Aristotle's own hierarchy, the nesting provides only a partial ordering. For some pairs of objects this criterion yields no decision as to which is more abstract. If the description of one object employs both form and function and the second only function or only causal structure, the first is more concrete than the second. But if one object is given just a causal structure and the second just a function, the criterion gives no verdict.

Consider a simple example. We might describe a paper-cutter as an implement which rends paper. The definition refers just to the function of paper-cutters. How a paper-cutter rends paper is not a factor in the explanation. But the definition of scissors must give an account of the parts of scissors and how they cut, e.g. that they have at least two blades which operate by sliding past each other. We do not need to specify the material or constituents of scissors. Any material from which blades can be made will do. So too with the materials of lasers. Yet scissors are more concrete than paper-cutters because more is specified about how the object works and performs its function. Note that we could specify functions more finely. A smooth cutter will create smooth edges in rending paper. Smooth cutters are more abstract than scissors. For the variety of objects which can be scissors is restricted in a way in which smooth cutters are not. Their structural and material possibilities are still more diverse.

It might be objected that this notion of the abstract/concrete distinction is merely a variety of the general/specific distinction. But this is only partly the case. It is true that scissors are kinds of cutter, but it is not true that scissors are kinds of paper-cutter, and much less that they are kinds of smooth paper-cutter—some scissors cut meat. Yet we claim that paper-cutters are more abstract than scissors.

A second objection is that one could very vaguely indicate a structure and function for one item and specify a very specific function for another. For example, consider a mover where right and left are distinguished and an implement which cuts paper into frilled, square mats. Surely the cutter of frilled, square, paper mats is more concrete than the right–left mover, and surely this shows that we really ought to have given a general/specific distinction. What the example shows, rather, is that the general/specific distinction is relevant, and that our notion presupposes that the levels of description be not too divergent.

It is easy to see why this should be from the viewpoint of our distinction. As Aristotle observes about structure in the *Physics*, one cannot have just any matter with a given structure, but structure presupposes certain sorts of matter (one cannot make a laser of any material). So too a function presupposes certain structures (reservoirs have to be big). If the specification of structure is too general, as in the case of right–left movers, and we compare it with a function as specific as that of cutters of frilled, square, paper mats, we may get the impression that the very general structure and function is less specific than the highly specific function and structures presupposed by it. In other words, the class of structures presupposed by the functional description of cutters of frilled, square, paper mats may be more specific than the structure mentioned in the description of right–left movers.

Let us now apply this criterion to the example of the laser and the reservoir. It is one of the corollaries to my doctrines on capacities that covering-law accounts are only one special form of explanation offered in contemporary science. Both the laser and the reservoir are good examples. In both cases it is at least as important to the model to specify the capacities of the systems modelled as it is to give the equations which describe their behaviour. The two cases differ in how much detail is provided about how the capacities operate. As I argued above, in characterizing the laser we specify its complex causal structure and we specify what it does; in the case of a reservoir we specify only what it does. This is typical in modern science: explanatory models differ in what kinds of explanatory factor they fill in. The schema that Mendell and I suggest for categorizing the various kinds of explanatory factor that may be involved is taken very roughly from Aristotle's system of the four causes. Of course Aristotle's notion of perfection and function is in no way limited to what a system does, just as form is not precisely the causal structure. The perfection in Aristotle is sometimes the proper activity or state of a system, sometimes the proper result of a process; the form is usually the structure of the working system, which is sometimes identified with the activity itself. Here, inspired by Aristotle's distinction , we propose to treat *form* analogously as how the system operates and *function* as what it does. We may say, then, that lasers are more concrete than reservoirs because they are specified in terms of both form and function, whereas reservoirs are specified in terms of function alone.

It is easiest to illustrate our concept of abstraction in the context of Aristotle's division of explanatory factors into four kinds; but this particular division is not necessary. Our basic view is that one system is more concrete than a second when the kinds of explanatory feature used to identify the second are nested in those that identify the first. So the general strategy requires only that we have a way of sorting explanatory features from non-explanatory features, and of determining what kinds of explanatory feature there are. It is a common—though we think mistaken—assumption about modern physics, for example, that function is not an explanatory feature at all. If so, then structure and function is no more concrete than a structure alone.

Even when a number of different kinds of explanatory feature are admitted by a science, as with Aristotle's four causes, not all need be seen as equally central. McMullin, we have seen, argues that the causal structure is the primary explanatory feature in modern physics, and we agree. For Aristotle, too, structure plays a privileged role. It is the most important of the four explanatory factors in physics. The constituent is what a substance is made of; the function what it does or becomes; the source where it comes from. All these are a part of physical explanation. But structure is what the substance is. It is the essence of the substance, or what makes this material body what it really is. Structure, therefore, among all four of Aristotle's causes, gives most fundamentally what a thing is, and hence by our criterion contributes most to making it concrete. The one factor that may seem equally important is matter.[52] For there is a

---

[52] Here a more sophisticated view of matter may be appropriate. In his last thoughts on matter (Met. Z10–11), Aristotle wondered if the parts of the substance which occur in the definition which gives the essence could be matter. Matter as the locus of possibility and accident cannot occur in the essence or its definition. Hence, anything which appears in the definition which gives the essence, even a constituent of the substance, cannot be matter. We have here a way of distinguishing the Duhemian abstract and concrete. Matter will appear in the account of the concrete objects, but it cannot appear in the account of the abstract object. For that will only include features essential for the particular theory. On the other hand, we should not think that via Aristotle we can capture the notion of the Duhemian abstract object. For Aristotelian form or structure will be the structure of the concrete object, while the Duhemian abstract structure will not. In this respect Duhem is a Platonist.

sense in which matter—as that which individuates structure—is more concrete: it serves more to isolate the individual. But this is a different sense from the one we have been considering. For designating or picking out the individual does not tell what it is. This is primarily the job of the structure.

In the case of lasers and reservoirs, the two relevant explanatory factors that we have isolated are function and structure. Between these two factors there is yet another reason for giving priority to causal structure in determining concreteness. For, given the causal structure, the function often comes along as a by-product, but not vice versa. One cannot usually infer from effects to a causal history. At best one can describe classes of structures which would have such effects, only one of which could be the structure of the particular system at hand. But one can, at least in principle, infer from a description of the causal structure to what that causal structure does. This is not to say that one may not also infer all sorts of side-effects. But all these effects are effects of that system as described. If a system is described in terms of causal structure, the addition of a description of certain effects of the system as essential serves merely to isolate those effects we are interested in from all the actual effects of the system. For this reason, descriptions of effects seem to add less to the descriptions of well-specified causal structures than do descriptions of causal structure to descriptions of well-specified causal effects. Hence, descriptions of specific causal structure are, in general, more concrete than descriptions of specific causal effects.

We see, therefore, that causal structure is privileged for a variety of reasons. From either our point of view, or the more traditional Aristotelian one, causal structure plays the central explanatory role. It thus contributes most significantly to removing an object from the realm of the abstract, and making it more concrete. Consider the example of the crystal diode rectifier mentioned at the beginning. We can isolate three levels of concreteness here. The rectifier is identified by function alone; the diode rectifier by function and structure; and the crystal diode rectifier by function, structure, and matter.[53] Each is more concrete than the one before; but, we find intuitively, the first step, where causal structure is added to function,

---

[53] Van Nostrand's *Scientific Encyclopedia* (Princeton, NJ, 1958), 519, characterizes a crystal diode as 'a diode consisting of a semiconducting material such as germanium or silicon, as one electrode, and a fine wire "whisker" resting on the semiconductor as the other electrode. Because of its low capacitance, the device finds considerable application as a rectifier or detector of microwave frequencies.'

is far greater for purposes of scientific understanding than the second, where the material is filled in. This mirrors the view of section 5.4.3 that material abstraction in science works by subtracting all those factors which are only locally relevant to the effect. What is left then is just factors which will be relevant independent of context, and these seem to be the proper subject of scientific understanding. This contrast between the rectifier and the diode rectifier parallels the case of the laser and the reservoir. Adding causal structure to functional description, as in a laser, produces a critical change in the degree of concreteness of the object. We have seen this critical difference in the contrast between the laser and the reservoir.

Aristotle did not treat his four kinds of explanatory factor on a par. Structure or form supplies the essence, what the substance most truly is. If Mendell and I are right, modern science also gives central place to structural factors. Given the Aristotelian ideal that the explanatory factors are what tell us what an object really is, an object without structure—like a reservoir—will be a shadow object indeed.

5.6. Conclusion

Mill used the term 'tendency' to refer to what the laws of nature are about. My views are almost identical to his and yet I use the term

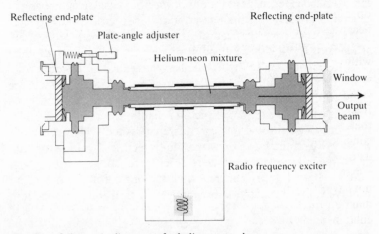

FIG. 5.1   *Schematic diagram of a helium-neon laser*
Source: A. L. Schawlow, 'Optical Masers', *Scientific American*, 204 (June 1967), 135.

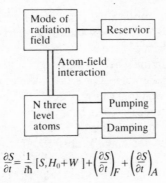

$$\frac{\partial S}{\partial t} = \frac{1}{i\hbar} [S, H_0 + W] + \left(\frac{\partial S}{\partial t}\right)_F + \left(\frac{\partial S}{\partial t}\right)_A$$

FIG. 5.2 *Block diagram of a laser*
Source: W. H. Louisell, *Quantum Statistical Properties of Radiation* (New York: Wiley, 1973), 470.

'capacity'. Why? The answer to this seemingly trivial terminological point provides a last argument in favour of capacities. The familiar example of lasers will illustrate. Consider Figs. 5.1 and 5.2. Fig. 5.1 is a schematic diagram of a laser constructed from specified materials in a specified way—it is a diagram of the first helium–neon gas laser. The diagram leaves out a large number of features, possibly even a large number that are relevant to its operation. Let us grant for the sake of argument that the concrete object pictured here operates under some law. I call the law a phenomenological law because, if we could write it down, it would literally describe the features of this concrete phenomenon and the nomological links within it. It would be a highly complex law, and would include a specific description of the physical structure and surroundings of the concrete device. Possibly it could not be written down; perhaps the features that are relevant to its operation make an open-ended list that cannot be completed. But because we are talking about a concrete object, it is at least conceivable that some such law is true of its operation.

Now look at Fig. 5.2 and its accompanying equation. Here we do not have a schematic diagram for a particular laser but rather a block diagram for any laser. In the block diagram we consider no particular medium, no specific way of pumping, no determinate arrangement of the parts. We abstract from the concrete details to focus, as we say, 'on the underlying principles'; the equation written beneath it records one of these principles. I will not explain the equation; it is not important to the philosophical point to know exactly what it says. The important question is: 'How do these two diagrams differ?' In the language of the last two sections, the second is a material abstraction whereas the first is not. The first diagram is a representation and not a real laser, so it will of necessity be incomplete; we have allowed that this may also be true of any equation we try to write down to describe its behaviour: no matter how long an equation we write for the laser of Fig. 5.1, it may not be literally true. But that does not make it abstract. The second equation is abstract; and that is why it is so short. It is not even an attempt to describe the actual behaviour of any material laser. Rather, it tells what a laser does by virtue of its function and the arrangement of its parts.

Here again I use the Aristotelian concepts which seem to me the natural way to describe the explanatory structures in physics. Again I will postpone past the end of this book the question of what this signifies about nature or about our image of it. The point I do want to make here is that no causality seems to be involved in either Fig. 5.1 or Fig. 5.2, nor in the equations appropriate to them. The example of an abstract law from laser theory in section 5.4 was different: 'An inverted population can lase.' That law tells what an inversion tends to cause. But the equation that accompanies the block diagram does not say what the laser tends to cause; rather, it says what the laser tends to do, or how it tends to behave. Here I think a difference in terminology is appropriate. Properties may carry a variety of different kinds of tendency—tendencies to behave in certain ways, to bear certain fixed relations to other properties, to evolve in a particular manner, or to produce certain kinds of effects. For all of these I use the general word 'tendency'; 'capacity' is reserved for a special subset of these—those tendencies which are tendencies to cause or to bring about something.

This book has been concerned with causation, and so has focused on capacities. But what about the broader category: are more general kinds of tendency real? My remarks in this conclusion are speculative and forward-looking, to projects yet to be undertaken; but it seems to me that a careful study will show that more general kinds of tendency *are* real; for not many of the arguments I have given in favour of capacities rely on their peculiarly causal nature. Following Mill, I have spoken of the composition of causes. But it seems fairly clear that, in giving scientific explanations and in making scientific predictions, we compose not only causes but behaviours.

This is transparent in physics, where equations and formulae are involved. The model Hamiltonians of section 5.4 are a good example. Although I introduced them into the middle of a discussion of capacities, there is nothing especially causal about them: to say, 'The Hamiltonian for a hydrogen atom is –' is not to say what the hydrogen atom produces but rather what energy configurations are possible for it. Yet Hamiltonians compose, just as capacities do; and if we do not think that the laws which describe the Hamiltonians for different kinds of situation are laws ascribing tendencies to these situations, how shall we interpret them to make them true? Read as pure statements of regularity, they are riddled with exceptions and seem to face all the same problems that led Mill to adopt tendencies. Indeed, although I put Mill to use to defend my doctrines in favour of stable causal tendencies, I think Mill himself had no special concerns with causality, and aptly used the more general term 'tendency' rather than my narrower conception of 'capacity'.

A second major argument for capacities concerned the exportability of information: we gather information in one set of circumstances but expect to use it in circumstances that are quite different. Again, there need be nothing especially causal about the information. Section 5.2 argued that this kind of stability and exportability is necessary to make sense of our use of what are called 'Galilean methods'. These methods are in no way confined to the study of causes, but are at work in almost every empirical enquiry in modern science. A third defence for capacities involved a counter-attack against the proposal to modalize them away. Again, the concepts of interaction and duality of capacity that played a central role in that counter-attack are just as relevant whether it is capacities in particular that are in question or some more general notion of tendency, though of course where probabilistic causes are not involved the statistical characterizations of these concepts will be irrelevant. Also, the same circle of irreducible concepts of enabling conditions, preventatives, precipitators, and the like will be just as important for the proper statement of general tendency claims as they are for claims about capacities.

From a crude first look, then, it seems that most of the arguments for capacities will double as arguments for more general kinds of tendency, and be equally compelling in both cases. That in itself turns into a powerful argument against the Humean and in favour of capacities. In Chapter 4 I argued that the most general kinds of causal claim in science are at base claims about capacities. This chapter has proposed a different but not incompatible way of looking at them: they are also material abstractions. But recall the block diagram of the laser (Fig. 5.2). If the arguments just sketched turn out to work, the accompanying equation ought to describe a general tendency, a tendency that lasers have by virtue of the abstract structure pictured in the diagram. Equally, both the diagram and the equation can be viewed as material abstractions. The equation is not some kind of regularity law with *ceteris paribus* conditions repressed. The regularities appear in the phenomenal content, in the detailed equations that attempt to describe the behaviour of concrete lasers and of specific materials. Like capacity ascriptions, the abstract equation also has the troubling feature that there is no rule that takes you from the abstract law to the phenomenal content; the additions and corrections will depend in each instance on how the block structure is realized. But there is an important difference. In the case of a capacity claim, the realization-dependent laws that fall under it will be some complicated sort of causal law, whereas in general the content of an abstract claim in physics, like the equation for the block laser, may be entirely associational.

Return for a moment to my conclusions about causality. In the end I present a structure with three tiers. At the bottom we have individual singular causings. At the top we have general causal claims, which I render as statements associating capacities with properties—'Cash-in-pocket can inhibit recidivism', or 'Inversions have the capacity to amplify'. In between stands the phenomenal content of the capacity claim—a vast matrix of detailed, complicated, causal laws. These causal laws, which stand at level two, provide no new obstacles for the Humean beyond those that already appear at the bottom level. For these are just laws, universal or probabilistic, about what singular causings occur in what circumstances what percentage of the time. This claim was defended in Chapter 3.

Yet there is good reason for the Humean to be wary, at least the modern-day Humean who believes in laws but not in causes. For if the programme of section 4.6 succeeds, this treasured layer of laws will be eliminated altogether from our fundamental scientific picture. All that remains would be capacities and the causings for which they are responsible. The laws—even causal laws—would no longer play any basic role; and the pretence that nature is full of regularities could be dropped. This radical programme gets some support from the discussions of this chapter. For I hope it becomes clear that what are taken as the fundamental laws of, say, physics are not laws at all in the empiricist's sense, but abstractions. Section 5.5 proposes an Aristotelian scheme for treating the structure of abstractions in physics. This may not be the best way to do it. But whichever route is taken, it seems clear that the conventional idea of laws will not do at all. The abstract claims which physical theory consists of do not describe regularities, neither regularities about individual causal behaviour, nor regularities of association. What regularities there are occur at a far lower level in science, long, complicated, and local. They fit nobody's idea of a fundamental law.

But this is a programme. I hope to eliminate the middle level, but for the moment I have argued for a picture of science with three layers. The third is the most troubling of all for a conventional empiricist. It sounds doubly non-Humean. First we have causal connections in the single case; then we add some mysterious capacities, or occult powers. But the point of the block diagram is to show that the mystery concerning capacities is not isolated in the domain of causality, but spreads everywhere in physics. Sections 5.1–4 teach that a capacity claim, considered as an abstraction, has two distinct facets. First, it is a material abstraction, just like the equation for the block diagram and like most general claims in physics. Second, unlike the equation for the block diagram, the laws which make up the phenomenal content of the capacity claim are causal laws, not equations or other laws of association. The use of the term 'capacity law' rather than the more general 'tendency law' marks this special feature of the phenomenal content.

Yet neither facet introduces any new troubling ideas that are peculiar to causal discourse, beyond the already necessary notion of singular causation. What is especially causal about a capacity claim are the causal laws that make up its phenomenal content; these have just been discussed. Beyond that there is only the fact of what I have called 'material abstraction', and material abstraction cuts across the causal and the non-causal in science: whether causal or not, material abstractions cover a vast network of concrete laws which are both complex and detailed; and whether causal or not, the form of the concrete laws is highly realization-dependent. I have claimed that we have not even a good start on a philosophical account of how this realization-dependence works. But that is not any special problem to do with causality, since the process of material abstraction is fundamental throughout physics and probably to all of modern theoretical science. So the picture I draw of singular causings, causal laws, and capacities is at least not doubly non-Humean. Once we let in singular causation—as we must—we have no more special problems with capacity ascriptions than with any other abstract descriptions in physics. The most general claims of physics are indeed mysterious; but the causal claims are not more mysterious than others.

6

What Econometrics Can Teach Quantum Physics: Causality and the Bell Inequality

6.1. Introduction

At least since the 1870s social scientists have been prone to take physical theory as their model, and this has been no less true in economics than elsewhere. Adolphe Quételet's ideas for a social physics are one good example. A recent history[1] even claims that the probabilistic methods of econometrics emerged in an attempt by neo-classical economists to keep up with the indeterminism of the newly developed quantum theory. But there is at least one case in which the traditional order can be reversed, and in which the methods of econometrics can be used to disentangle a problem in quantum mechanics. This is the case of Bell's inequalities.

Bell's inequalities are supposed to show that no local hidden-variable theories are possible in quantum mechanics. But 'local' here is a peculiar term. It names a statistical condition that looks much like the common-cause conditions from Herbert Simon, described in section 1.2. The alternative name sometimes used, 'factorizability', is closer to what the statistical condition actually requires. The condition is essential to the derivation of the no-hidden-variables result, and therefore it is important to understand exactly what it means. This is where the structural models of econometrics help: the conclusions about structural models arrived at in Chapter 3 show that factorizability is not, as many assume, a criterion for the existence of a common cause.

Bits of this chapter are technically complicated. But the ideas are

[1] P. Mirowski, *More Heat Than Light: Economics as Social Physics* (Cambridge: Cambridge University Press, 1988). This view is not uncontroversial. For an alternative account see M. Morgan, 'Statistics Without Probability, and Haavelmo's Revolution in Econometrics', in L. Krüger, G. Gigerenzer, and M. Morgan (eds.), *The Probabilistic Revolution: Ideas in the Social Sciences* (Cambridge, Mass.: MIT Press, 1987).

not. Readers who are not especially interested in the details should be able to follow the main arguments without attending to the equations. A review of section 3.3.3 should help as well.

6.2. Bell's Inequality

The inequalities were first derived by J. S. Bell in 1964. Since then there have been a number of different versions of them, based on Bell's methods, as well as a number of experiments confirming quantum mechanics at the expense of hidden variables. The setting is an experiment originally proposed by Einstein, Podolsky, and Rosen, in a paper which argues in favour of hidden variables and maintains that quantum mechanics is incomplete because it fails to describe them. I will call the experiment the EPR experiment, even though what I shall describe is a more modern version, more simple than the original proposal of Einstein, Podolsky, and Rosen and abstracted from any real experimental complications.[2]

The experiment measures the spin components at two different angles, θ and θ', of two spin $- \frac{1}{2}$ particles widely separated and travelling in opposite directions. The particles might be electrons. To say that they are spin $- \frac{1}{2}$ particles means that, regardless of the angle at which they are measured, they will be found to have a spin component of either $+ \frac{1}{2}$ (called 'spin-up') along that angle, or of $- \frac{1}{2}$ ('spin-down'). The particles are supposed to have been emitted from a source with no angular momentum. Quantum mechanically they are said to be in 'the singlet state'. Since angular momentum is conserved in quantum mechanics, and spin is a kind of angular momentum, one particle will yield spin-up for a given angle if and only if the other yields spin-down. Quantum mechanics predicts that the outcomes will also be correlated for measurements along different angles. Let $x_L(\theta)$ represent the proposition, 'the outcome is $+ \frac{1}{2}$ in a measurement along angle θ on the particle in the left wing'; similarly for $x_R(\theta)$, in the right wing. The quantum mechanical predictions are that

---

[2] For an account of the contents and history of the EPR paper, see A. Fine, *The Shakey Game* (Chicago, Ill.: Chicago University Press, 1987). For more details about the Bell inequalities and their experimental testing, as well as for further references, see M. Redhead, *Incompleteness, Nonlocality, and Realism* (Oxford: Clarendon Press, 1987). The Einstein *et al.* paper was published in *Physical Review*, 46 (1935), 777–84.

$$P(x_L(\theta).x_R(\theta')) = \tfrac{1}{2}\sin^2\{(\theta - \theta')/2\} = P(\neg x_L(\theta). \neg x_R(\theta'))$$
$$P(x_L(\theta). \neg x_R(\theta')) = \tfrac{1}{2} - \tfrac{1}{2}\sin^2\{(\theta - \theta')/2\} = P(\neg x_L(\theta).x_R(\theta'))$$

Here the dot (.) indicates logical 'and'.

These correlations between distant outcomes are the source of the puzzle. The simplest account of the correlations would be that each pair of particles was supplied at the source with a spin component in every direction, where the particle on the left got spin-up for a given angle just in case the right-hand particle got spin-down. This is the account urged by Einstein, Podolsky, and Rosen. But spins in different directions are incompatible quantities in quantum mechanics. That is typically taken to mean that they cannot have well-defined values at once (the more familiar case is position and momentum). Indeed, if a system can have well-defined values for spin along different directions, this is a fact that no state in quantum mechanics can express, and that is why Einstein, Podolsky, and Rosen accused the theory of incompleteness.

Bell is supposed to have shown that the ordinary understanding of quantum mechanics is right. The correlated outcomes cannot be accounted for by assuming that there are values there to be measured all along; even more strongly, no hidden or unknown variables at all can be postulated to account for the correlations without allowing mysterious collusion between the separated systems at the instant of measurement. The collusion would be mysterious because the two systems are supposed to be measured simultaneously, with no time to communicate to each other what response they will make to the apparatus they see. Hidden-variable theories without such collusion are called 'local' theories. What Bell did was to derive a set of inequalities that any local hidden-variable theory must satisfy in an EPR experiment. Quantum mechanics does not satisfy these inequalities, and it is quantum mechanics that has been subsequently experimentally confirmed.

Locality is expressed in a familiar-looking factorizability condition. Labelling the hidden variable x_1, Bell's derivation assumes

Factorizability in EPR

$$P(x_L(\theta).x_R(\theta')/x_1) = P(x_L(\theta)/x_1)P(x_R(\theta')/x_1)$$

A superficial comparison with Simon's conditions may make factorizability appear the perfect way to express the fact that x_L and x_R are not supposed to influence each other. For it is exactly analogous to the factorizability condition of section 1.2. But this is not the case. Section 3.3 showed what the general criterion for a common cause in a three-variable model is. Simon's is a special case and one not suited to EPR.

6.3. A General Common-Cause Criterion for the EPR Experiment

A simple example will illustrate how Simon's structure goes wrong. It is an example similar to ones proposed by Bas van Fraassen in criticisms he makes of Wesley Salmon's use of factorizability as a common-cause condition.[3] Suppose that a particle collides with an atom and the atom emits two new particles as a consequence. Particle 1 may be emitted either at angle θ or at angle θ'; and the probabilities are 50–50. Particle 2 may be emitted at $-\theta$ or at $-\theta'$; and again the probabilities are 50–50. But momentum must be conserved. So particle 1 is emitted at θ if and only if particle 2 follows the path defined by $-\theta$.

It is obvious in this situation that a cause may be at work: λ, when it is present in the atom, produces motions along θ, $-\theta$; otherwise the atom emits the particles at angles θ' and $-\theta'$. In this example λ may be a deterministic cause: P(the angle of emission for particle $1 = \theta/\lambda) = 1 = $ P(the angle of emission for particle $2 = -\theta/\lambda)$, with $P(\lambda) = .5$. Or it may be a purely probabilistic cause: P(the angle of emission for particle $1 = \theta/\lambda) = r = $ P(the angle of emission for particle $2 = -\theta/\lambda)$; in which case, since $\frac{1}{2} = P(\theta.\lambda) + P(\theta.-\lambda) = P(\theta.\lambda)$, $P(\lambda) = 1/(2r)$. If λ is totally deterministic in producing θ and $-\theta$, the probabilities will factor: P(the angle of emission for particle $1 = \theta$ and the angle of emission for particle $2 = -\theta/\lambda) = 1 = $ P(the angle of emission for particle $1 = \theta/\lambda) \times $ P(the angle of emission for particle $2 = -\theta/\lambda) = 1 \times 1$. But if the probabilities depart from 1 at all, factorizability will no longer obtain: P(the angle of emission for particle $1 = \theta$ and the angle of emission for particle $2 = -\theta/\lambda) = r \neq $ P(the angle of emission for particle $1 = \theta/\lambda) \times $ P(the angle of emission for particle $2 = -\theta/\lambda) = r^2$.

But in this case it is not reasonable to expect the probabilities to

---

[3] B. van Fraassen, 'Rational Belief and the Common-Cause Principle', in R. McLaughlin, *What? Where? When? Why?* (Hingham, Mass.: Kluwer Boston, 1982).

factor, conditional on the common cause. Momentum is to be conserved, so the cause produces its effects in pairs. This is just a special case of a cause which operates subject to constraints, a case familiar from section 3.3. In this example, whether one of the effects is produced or not has a bearing on whether the cause will produce the other effect. But factorizability is a criterion tailored to cases where the cause operates independently in producing each of its effects. Clearly it is not an appropriate condition for this example.

Before discussing the EPR experiment in more detail, it will be useful to translate the equations of Chapter 1, and the probabilistic criteria they give rise to, out of the language of random variables and into that of Boolean logic. This will provide immediate contact with the Bell literature, where it is customary to discuss the experiments using indicator functions; that is, functions which take the value 1 if the outcome in question—here, spin-up—occurs, and 0 if it does not. This demands modification in the shape of the equations, since no matter how many of the causes occur, the effect variable will never get larger than 1. One strategy to deal with this is to keep the random-variable format, but add terms to the equations. For illustration, the analogue to the three-variable equation of Chapter 1, for an effect x_3 whose causes are supposed to be x_1 and x_2, is

$$x_3 = \hat{a}_{31}x_1 + \hat{a}_{32}x_2 + u_3 - \hat{a}_{31}x_1u_3 - \hat{a}_{32}x_2u_3 - \hat{a}_{31}x_1\hat{a}_{32}x_2 \\ + \hat{a}_{31}x_1\hat{a}_{32}x_2u_3$$

where x_1, x_2, x_3, and u_3, like $\hat{a}_{31}$ and $\hat{a}_{32}$, are allowed only the values 0 and 1.

The extension to more exogenous variables is straightforward. Alternatively, one can take the indicator functions to represent yes–no propositions, and use the notation of Boolean logic:

$$x_3 \equiv \hat{a}_{31}{\cdot}x_1 \ v \ \hat{a}_{32}{\cdot}x_2 \ v \ u_3$$

In that case the 'equation' will look just like those in Mackie's *inus* account of causation discussed in section 1.3. The two strategies are equivalent. I shall use the second because it is notationally simpler.

Letting x_1, x_2, x_3, represent yes–no questions about events at $t_1 < t_2 < t_3$ which may be causally linked, the three-variable structure will look like this:

$$x_1 \equiv u_1$$
$$x_2 \equiv \hat{a}{\cdot}x_1 \ v \ u_2$$
$$x_3 \equiv \hat{b}{\cdot}x_1 \ v \ \hat{c}{\cdot}x_2 \ v \ u_3$$

The derivation of a common-cause condition proceeds in exact parallel with Chapter 1. The introduction of $\hat{a}$, and $\hat{b}$, and $\hat{c}$ allows the structure to represent causes that work purely probabilistically: $\hat{a}$ represents the event of x_1 operating to produce x_2; $\hat{b}$, x_1's operation to produce x_3; and $\hat{c}$, x_2's operation to produce x_3. In keeping with this interpretation, it will be assumed that the operations of distinct causes are independent of each other and of everything that occurs previous to them. As in the random-variable structures, the us may be necessary to ensure identifiability; but in general they are not, and can often be omitted.

So, dropping terms in the us, it follows that

General common-cause criterion
$$\hat{c} = \perp \text{ iff } P(x_2 \cdot x_3/x_1) = P(\hat{a} \cdot \hat{b}/x_1)$$

where $\perp$ indicates 'zero' or logically false. This is true so long as $x_1 \not\rightarrow (\hat{a} \rightarrow \hat{b})$, i.e. so long as x_2 is indeed a cause of $x_3 (\hat{b} \neq \perp)$, and sometimes x_1 produces x_2 without producing x_3. The condition says that x_2 plays no role in producing x_3 if and only if the joint occurrence of x_2 and x_3 is the same as the probability for x_1 to produce them jointly (plus, of course, any contributions from the omitted error terms).

To get from this general common-cause condition to factorizability, one sets

$$P(\hat{a} \cdot \hat{b}/x_1) = P(\hat{a}/x_1)P(\hat{b}/x_1)$$

from which follows

Factorizability
$$P(x_2 \cdot x_3/x_1) = P(x_2/x_1)P(x_3/x_1)$$

Thus the factorizability condition does indeed depend on the assumption that the joint cause acts independently in producing its separate effects. This assumption does not fit the example above when the cause produces its outcomes in pairs. For that case the more general formula must be used. Clearly, one cannot just assume that factorizability is the right common-cause condition in EPR either.

6.4. Quantum Realism and the Factorizability Condition

There is a large literature discussing whether factorizability is an appropriate constraint to put on a hidden-variable theory for EPR, but before approaching that literature one needs to have a clear idea what purposes a hidden-variable theory is supposed to serve. Why want hidden variables? There are two distinct answers: to restore realism in quantum mechanics; and to restore causality. It turns out that factorizability is a very different condition in the two settings. This section will discuss the first answer.

Quantum systems notoriously are not supposed to have all the properties they should at any one time. Position and momentum are the classic examples; as already mentioned, spin components in different directions are also not simultaneously well defined. Quantum mechanics' failure to treat classic dynamic quantities, like position, momentum, energy, and angular momentum, as real all at once has been a major problem since the early days of the theory. It is well known that Einstein was deeply troubled by this aspect of quantum mechanics; his concern is reflected in the EPR paper. He thought that objects (at least macroscopic objects) do have all their dynamic properties at the same time. Quantum mechanics does not allow this; it is hence incomplete.

But is it true that realism of this kind is incompatible with quantum mechanics? J. S. Bell claims[4] that, until the time of his own work, Einstein might have been right. But the Bell inequalities, and their subsequent experimental tests, show that he was not. This claim of Bell's clearly relates the inequalities to the realism issue: the hidden variables are just the values that the spins really have, 'hidden' because they cannot all be measured at once.

In the context of quantum realism, factorizability is immediate:

$$P(A \cdot B / A \cdot B) = 1 = P(A / A \cdot B)P(B / A \cdot B) \text{ and } P(A \cdot B / \neg A \cdot B) =$$
$$0 = P(A / \neg A \cdot B)P(B / \neg A \cdot B) = 0 \times 1$$

Given the values for the spin, all the probabilities are one if the selected outcomes match the values; otherwise the joint probability is zero, and so too is at least one of the two marginal probabilities on the right. The rationale for the factorizability condition is thus entirely trivial. The same will not be true when the question is one of causality.

Even here, in the context of realism, there are problems with the condition, for the simple argument above about zero–one probabilities hides a number of non-trivial assumptions. One will do for illustration. The probabilities that matter in the Bell inequalities are probabilities for measurement outcomes. These are the probabilities that quantum mechanics predicts and that the various experiments can test. The argument for factorizability in the context of quantum realism assumes that, once all the normal corrections are made for experimental error, the measured values match the true values, and occur with the corresponding frequencies. This may not be the case. Of course, if the relation between the two is entirely unsystematic, the realism obtained would be vacuous. But Arthur Fine[5] has produced a number of models in which measured and possessed values do not match, and in which the assumption of such a match fails for specific reasons which are different in each model. In each case the factorizability condition is no longer appropriate. In a rigorous derivation of factorizability in the context of quantum realism, all the hidden assumptions, like those pointed out by Fine, would have to be laid out. But the point here is the converse: violations of the assumptions may provide reasons to reject factorizability; the reason to accept it as a requirement for quantum realism is just the trivial argument based on zero–one probabilities.

The second reason to look for hidden variables is to restore causality in quantum mechanics. Often the two reasons are not carefully distinguished. This is probably because, for many, an interest in causality is tantamount to an interest in determinism, and factorizability looks the same from a deterministic point of view as from a realistic one. Just partition the set of hidden states into classes: $x_1(x_L(\theta), x_R(\theta'))$ labels all the states that produce spin-up on the left in a measurement along θ and spin-up on the right along θ'; $x_1(\neg x_L(\theta), x_R(\theta'))$ labels those that give spin-down on the left along θ and up on the right along θ', and so forth. In that case, for the purposes of the Bell derivation the hidden states might as well represent the possessed spins themselves, for factorizability follows in exactly the same way:

$$P(A \cdot B / x_1(A \cdot B)) = 1 = P(A / x_1(A \cdot B)) P(B / x_1(A \cdot B)), \text{ and}$$
$$P(\neg A \cdot B / x_1(A \cdot B)) = 0 = P(\neg A / x_1(A \cdot B)) P(B / x_1(A \cdot B)) =$$
$$0 \times 1$$

6.5. A Common-Cause Model for EPR

When the hidden causes act stochastically, the situation is entirely changed. Where, then, does factorizability come from? It is usual,

[5] 'Counting Frequencies', *Synthese*, 42 (1979), 145–54.

when applying a causal interpretation to the Bell results, to talk as if factorizability is a common-cause condition. But we have seen that this is not always appropriate. Different causal structures give rise to different common-cause conditions; the EPR set-up, it turns out, requires something other than factorizability. Indeed, the most primitive structure appropriate for EPR is similar to the one at the beginning of section 6.3. Conservation constraints apply. If there is a common cause in the source, it will not produce effects in the two wings independently. The operations for the two effects are correlated, and quantum mechanics tells us, essentially, what those correlations are.

Using the earlier example as a guide, it is easy to construct a simple common-cause model for EPR. A slightly more complicated model will be presented as well.

Common-Cause Model for EPR 1

$$x_L(\theta) \equiv \hat{a}_L(\theta){\cdot}x_1 \vee u_L(\theta) \tag{1}$$

$$x_R(\theta) \equiv \hat{a}_R(\theta){\cdot}x_1 \vee u_R(\theta) \tag{2}$$

$$P(x_1) = \tfrac{3}{4} \tag{3}$$

$$P(\hat{a}_L(\theta){\cdot}\hat{a}_R(\theta')) = \tfrac{2}{3}\sin^2\{(\theta - \theta')/2\} = P(\neg\,\hat{a}_L(\theta){\cdot}\neg\,\hat{a}_R(\theta')) \tag{4}$$

$$P(\hat{a}_L(\theta){\cdot}\neg\,\hat{a}_R(\theta')) = \tfrac{2}{3} - \tfrac{2}{3}\sin^2\{(\theta - \theta')/2\} = P(\neg\,\hat{a}_L(\theta){\cdot}\hat{a}_R(\theta') \tag{5}$$

The usual randomness assumptions on the us makes them independent of each other and of x_1 and its operations, in all combinations:

$$P(\pm x_1 \pm \hat{a}_L(\theta) \pm \hat{a}_R(\theta') \pm u_L(\theta) \pm u_R(\theta')) = P(\pm x_1)$$
$$P(\pm a_L(\theta) \pm \hat{a}_R(\theta'))P(\pm u_L(\theta))P(\pm u_R(\theta')) \tag{6}$$

Equation (6) might be viewed as a kind of locality constraint: the background production of 'up' outcomes in the left wing does not depend on the background in the right wing, and vice versa. It is not, however, the problematic assumption that usually goes by that name.

For symmetry, let

$$P(u_L(\theta)) = P(u_R(\theta')) \tag{7}$$

and assume

$$O < P(u_{L(R)}(\theta)) \leqslant \epsilon/(1 + 2P(\hat{a}_{L(R)}(\theta)) = \epsilon/(1 + 2(2/3)) = 3\epsilon/7 \tag{8}$$

where ϵ is the amount of experimental error admitted. Obviously the model satisfies the appropriate common-cause condition.

Common-Cause Condition 1

$$P(x_L(\theta) \cdot x_R(\theta')/x_1) = P(\hat{a}_L(\theta) \cdot \hat{a}_R(\theta')/x_1) + \ldots$$

since the equations are constructed both to make x_1 a common cause of x_L and x_R, and to make these two causally irrelevant to each other. The omitted terms in the Common-Cause Condition 1 are all functions of $P(u_L(\theta))$ and $P(u_R(\theta))$; equation (8) guarantees that the exact quantum prediction, which is on the right-hand side of (9), is within ϵ of the model prediction, on the left-hand side.[6]

One important feature must be noted that distinguishes this quantum model from the more familiar models of the social sciences. In the usual construction of structural models, one begins with a field of propositions generated from elements like x_1, $x_L(\theta)$, $x_L(\theta')$, $x_L(\theta'')$, ..., $x_R(\theta)$, $x_R(\theta')$, ..., and then prescribes a probability measure over this field. That is, one begins with a giant joint probability over the common cause and all its possible effects. But with EPR this will not be possible. A number of different proofs show that joint probabilities for incompatible outcomes like $x_L(\theta)$ and $x_L(\theta')$ are not permitted in quantum mechanics. The most recent proof is due to Arthur Fine, and is of immediate relevance here. Fine has shown[7] that there is a function suitable to serve as a single joint distribution for all the observables of an EPR-type experiment if and only if there is a factorizable, stochastic hidden-variables model, and hence if and only if the Bell inequalities are satisfied. So a model compatible with quantum mechanics, like the common-cause models here, where factorizability fails, cannot be one where the probabilities sit in a common-probability space. Instead, the

---

[6] In a three-variable common-cause model:

$$x_2 \equiv \hat{a}x_1 \vee u_2$$
$$x_3 \equiv \hat{b}x_1 \vee u_3$$
$$\text{Exp}(x_2 \cdot x_3/x_1) = \bar{g} + (\bar{b} - \bar{g})\bar{u}_2 + (\bar{a} - \bar{g})\bar{u}_3 - \bar{u}_2\bar{u}_3(\bar{a} + \bar{b} - \bar{g} - 1)$$

where $\bar{g} = \text{Exp}(\hat{a} \cdot \hat{b})$, $\bar{b} = \text{Exp}(\hat{b})$, $\bar{a} = \text{Exp}(\hat{a})$, and $\bar{u}_i = \text{Exp}(u_i)$. Hence when $\bar{u}_2 = \bar{u}_3 = \bar{u}$ and $\bar{a} = \bar{b}$, as in EPR model No. 1, Exp $(x_2 \cdot x_3/x_1) - \bar{g}/ \leqslant 2\bar{u} \; \bar{a} + \bar{u}^2(1 - \bar{a}) < 2\bar{u} \; \bar{a} + \bar{u}^2 \leqslant \bar{u}(1 + 2\bar{a})$. Alternatively, the random backgrounds could be dropped throughout. The only case where this raises problems of identification is when $\theta' = \theta$. In that case the quantum mechanical correlation between x_2 and x_3 is perfect; hence $\hat{a} = \hat{b}$, so that $\hat{c} = \perp$ can no longer be identified or determined from the probabilities. (Recall the remark immediately following the general common-cause criterion in section 6.3.) Still, even in this case, the *us* are not really necessary, since it seems entirely reasonable to extrapolate conclusions about the causal structure for $\theta_L = \theta_R$ from the structures that can be identified whenever $\theta_L \neq \theta_R$.

[7] 'Hidden Variables, Joint Probability, and the Bell Inequalities', *Physical Review Letters*, 48 (1982), 291–5.

probabilities must be defined separately over each Boolean subfield generated by the external causes of the model plus their co-realizable effects. In model 1 this means $\{x_1, x_L(\theta), x_R(\theta'), u_L(\theta), u_R(\theta')\}$ for any choice of θ and θ'. Similar care must be taken in defining the probability spaces for any model where the Bell inequality is violated; and the same applies in generalizing from yes–no propositions to random variables.

Model 1 is an especially simple structural model for EPR which includes a common cause. The simple model can be amended in a variety of ways to reflect a more complicated structure. It is now usual in the Bell literature to treat the influence of the measuring apparatuses explicitly, so it may serve as a useful illustration to construct a structural model that does this as well. Here I will just present the model. But in Appendix I, I show how this model can be constructed, using the procedures of Chapter 1, from the natural assumptions that one wants to make about the relations in an EPR-type set-up—most notably that neither the state of the apparatus in one wing at the time of measurement, nor the outcome there, can affect the outcome of a simultaneous measurement in the distant wing.

The construction of Appendix I is entirely straightforward. Nevertheless, I want to lay it out explicitly in order to recall the advantages of using structural methods. It is usual in the Bell literature to present and defend factorizability in some intuitive fashion, often by use of commonsense examples. Even when the attempt is made to trace the condition to more basic assumptions (as in the work of Jon Jarrett or Linda Wessels, which is discussed in Appendix II), the background assumptions about the causal structure of the EPR situation are still expressed directly as probabilistic conditions. Structural methods are more rigorous. All probabilistic conditions are derived from the structural equations, and the method provides a recipe for the exact expression of background causal information. The results of Chapter 1 then guarantee that the criteria derived provide a true representation of the hypothesized causal structure.

From Appendix II, the enlarged model looks like this:

$$x_L(\theta) \equiv \hat{a}_{L,\theta} \cdot x_1 \cdot \hat{m}_{L,\theta} \vee u_L \tag{1'}$$
$$x_R(\theta) \equiv \hat{a}_{R,\theta} \cdot x_1 \cdot \hat{m}_{R,\theta} \vee u_R \tag{2'}$$

where $\hat{m}_{L,\theta}$ designates the set of left-wing apparatus states that are positively relevant to spin-up there; and similarly for $\hat{m}_{R,\theta}$ in the right wing. (Throughout, $\hat{y}$ will designate the indicator function for y, which takes value 1 when y obtains, and 0 otherwise.) This structure is familiar from the discussion of Mackie in section 1.3: x_1 is a common partial cause of the outcomes in both the left wing and the right wing. The complete cause in a given wing also includes the state of the apparatus in that wing at the time of measurement. In this model there is no causal influence from one wing to the other, and that is apparent in the probabilities. The appropriate probabilistic condition to mark the absence of effects from x_L to x_R, or the reverse, in a modified three-variable model of this sort can be readily derived:

Common-Cause Condition 2

$$P(x_L(\theta) \cdot x_R(\theta') / \hat{m}_{L,\theta} \cdot \hat{m}_{R,\theta'} \cdot x_1) \approx P(\hat{a}_{L,\theta} \cdot \hat{a}_{R,\theta'} / x_1)$$

The only new substantive assumption required in the derivation is that the operations of x_1, which occur at the source, are independent of the states the apparatuses take on somewhere else at a later time. This is a special case of the requirement from Chapter 3 that a causal theory should be 'local'. I return to this sense of locality below.

As already mentioned, a number of derivations of Bell-like results do take explicit account of the causal role of the measuring apparatuses. A necessary condition in these derivations is a requirement known as 'full factorizability':

Full factorizability

$$P(x_L(\theta) \cdot x_R(\theta') / \hat{m}_{L,\theta} \cdot \hat{m}_{R,\theta'} \cdot x_1) = P(x_L(\theta) / \hat{m}_{L,\theta} \cdot x_1) \, P(x_R(\theta') / \hat{m}_{R,\theta'} \cdot x_1)$$

It is apparent that full factorizability follows only if one adds to Model 2 the assumption that $P(\hat{a}_{L,\theta} \cdot \hat{a}_{R,\theta'} / x_1) = P(\hat{a}_{L,\theta} / x_1) P(\hat{a}_{R,\theta'} / x_1)$, and that assumption, it has been seen, does not fit the EPR arrangement.

6.6. Quantum Mechanics and its Causal Structure

The probabilities in equations (4) and (5) of the common-cause models for EPR should seem familiar; they are directly proportional to the quantum mechanical probabilities. Indeed, this particular model was chosen just to make that point. In fact, if x_1 is replaced by quantum mechanics' own description for the EPR systems at the end of the interaction, the singlet state (call it ϕ_{EPR}), the quantum-mechanical predictions can be generated just by setting $P(\phi_{EPR}) = 1$ and replacing the coefficient ⅔ by ½ wherever it appears. Assuming that

$$0 < P(u_{L(R)}(\theta)\,) \leqslant \epsilon/2$$

the predictions of the model and those of quantum mechanics will be the same, within the bounds of the experimental error ϵ. It seems that the quantum state itself can be a common-cause in EPR. Hidden variables are not needed to 'restore' a causal structure to quantum mechanics—the causal structure is there already.

It is important to be clear exactly what this means. Recall the two different motivations for introducing hidden variables: to restore realism, and to restore causality. Common-cause models like Nos. 1 and 2 have no bearing on the first issue. Rather, they are designed to answer the question, 'Is it possible, so far as EPR shows, to postulate a causal structure consistent with the probabilities that quantum mechanics sets out?' The answer is yes, and this means that quantum mechanics is a very special kind of probabilistic theory. From the point of view of the probabilities, what a causal structure does is to impose a complicated set of constraints: in a theory without causality, any set of consistent probabilistic laws is possible; with causality, the pattern of probabilities is drastically restricted. Contrary to what might have been assumed, quantum probabilities are restricted in just the right way. Simple common-cause models, such as those in the last section, show that the pattern of quantum probabilities in EPR is exactly the familiar pattern associated with a common-cause structure, and that it is indeed a structure of just the kind one wants: the quantum state consequent on the interaction operates, in conjunction with the separated apparatuses, as a joint cause of the results in each wing, with no direct causal connection between one wing and the other. So far as the probabilities go, the quantum mechanics of EPR is itself a perfectly acceptable causal theory. The conclusion that it is not acceptable comes from mistakenly taking factorizability as the way to identify a common-cause.

6.7. Factorizability and the Propagation of Causes

Probabilities are not all there is to causality. Hume put two different kinds of demand on a causal theory, and modern accounts usually agree: the cause and effect must not only be regularly associated; they must be contiguous in space and time as well. The statistical criteria discussed in Chapters 1 and 3 are the modern-day equivalent of Hume's demand for regular association. What about contiguity? It turns out that factorizability finds its proper role here. Though factorizability is not a universal criterion for common-causes, it is a necessary condition for causal propagation. But just on account of this, I shall argue, it is not an appropriate condition to apply in quantum mechanics.

The treatment of propagation that I shall follow is due to Wesley Salmon.[8] Salmon's is an instructive case because his recent book on causal processes falls naturally into two quite separate parts, matching Hume's two different demands. The first part deals with the requirement that causes and effects be regularly associated; it lays down a number of statistical measures like those described in Chapters 1 and 3. The second part deals with causal processes, and treats the question of how a causal influence propagates. It assumes that a cause never operates across a space–time gap, but instead must travel from the cause to the effect along some continuous path. The two conditions are quite distinct, though they are not always clearly separated, even in Salmon's work.

Salmon calls his theory 'the "At–At" theory' because it assumes that the influence propagates from the initial cause to its effect simply by being *at* every point on a continuous trajectory in between. One important sign of propagation for Salmon is that the sequence of events connecting the cause with the effect be capable of conveying some kind of signal, or 'mark': it should be possible to insert a mark into the process at its origin and to observe the same mark—or some appropriately evolved version of it—at the effect. But the mark itself is just a symptom. What is crucial is that the capacity to produce the effect, or some feature that carries this capacity, should propagate on a path from the cause across the space–time gap. Salmon's account is metaphysically extremely simple, yet it is rich enough to show the connection between factorizability and propagation, so it will be used here.

For the case of quantum mechanics, it seems reasonable to assume that the influence must propagate with the system; that is, that the capacity-carrying feature can occur only in regions and at times

---

[8] *Scientific Explanation and the Causal Structure of the World* (Princeton, NJ: Princeton University Press, 1984).

where the probability for the system to appear is non-negligible. The basic idea involved in the requirement of spatio-temporal contiguity can then be formulated quite simply. The notation introduced so far is not well suited to dealing with continuous quantities and their intervals. Rather than introducing a more cumbersome notation, I will instead make use of the convenient fiction that time is discrete, and that contiguity requires that a cause operating at t produce its effect at the very next instant, $t + \Delta t$, and at the same place. In that case, following the lead of the at–at theory, Hume's second demand requires that a condition which I shall call 'propagation' be satisfied.

Propagation

x_c at $\vec{r},t$ causes (or is a partial cause of) x_e at $\vec{r},t'$ only if either
(a) there is a c such that x_c at $\vec{r},t$ causes (or is a partial cause of) c at $\vec{r},t + \Delta t$ and c at $r',t' - \Delta t$ causes (or is a partial cause of) x_e at $\vec{r}',t'$ and c (or an appropriately time-evolved version of c) exists at every point on some continuous trajectory between $\vec{r},t$ and $\vec{r}',t'$, where no region along the trajectory is quantum mechanically prohibited, or
(b) there is a chain of features beginning with x_c at $\vec{r},t$ and ending with x_e at $\vec{r}',t'$, where each causes (or is a partial cause of) the next later one and conditions like those in (a) are satisfied in between.

One further assumption is needed to derive factorizability in the EPR situation, and it is an assumption that does not follow directly from the demand for contiguity of cause and effect. This is an assumption about how causes operate: distant causes cannot act in coincidence; once two causes are separated, their operations must be independent of each other. This is a special kind of locality condition, which can be formulated simply enough for the purposes at hand.

Locality of operation

Let $\hat{a}_1$ represent the operation of c_1 at $\vec{r},t$ to produce an effect e_1, and $\hat{a}_2$ the operation of c_2 at $\vec{r}',t'$ to produce e_2. Then $P(\hat{a}_1 \cdot \hat{a}_2) = P(\hat{a}_1)P(\hat{a}_2)$ unless $\vec{r} = \vec{r}'$ and $t = t'$.

The motivation for this kind of condition must be something like Reichenbach's Principle, which motivates the search for causal structure in EPR in the first place: there are no distant correlations without causal explanations. In EPR, as in the classical example in section 6.3, the causes act subject to the constraint that energy be conserved. But conservation of energy is not a kind of force or cause itself. As Linda Wessels argues,

Certainly there are conservation laws in classical theory that restrict the relations among properties of distant bodies. Conservation of energy and momentum are obvious examples. But the total energy of two spatially separated bodies (in the absence of any other influence) is conserved either because the bodies exert forces on one another or, if no mutual interaction is present, because the energy of each body is conserved separately. Conservation of energy does not put any constraints on the total energy of the two that is not already required by the laws governing the evolution and interaction of the bodies individually and pairwise respectively.[9]

Hence there is the requirement for locality of operation.

Combining the requirements for propagation and locality of operation with those already assumed generates a variety of models for the EPR set-up. The simplest is one in which the capacity to produce spin-up propagates directly from the joint cause, x_1, to each wing of the experiment with no interactions along the way. The lessons for factorizability are the same whether the simplest model is assumed or a more complicated one. As before, the model is built with the assumption that the state and setting of the apparatus ($m_L(\theta)$ or $m_R(\theta')$) in one wing is not a cause of the outcome in the other; and also, in keeping with the requirement for spatio-temporal contiguity of cause and effect, with the assumption that the capacity propagated to one wing ($c_L(\theta)$ or $c_R(\theta')$) has no effect on the outcome in the other. The equations for the most simple common-cause model with propagation are then:

$$x_L\theta \equiv \hat{a}_L(\theta)\cdot\hat{c}_L(\theta)\cdot\hat{m}_L(\theta) \; v \; u_L \tag{1}$$
$$x_R(\theta') \equiv \hat{a}_R(\theta')\cdot\hat{c}_R(\theta')\cdot\hat{m}_R(\theta') \; v \; u_R \tag{2}$$
$$c_L(\theta) \equiv \hat{b}_L(\theta)\cdot x_I(\theta) \; v \; v_L \tag{3}$$
$$c_R(\theta') \equiv \hat{b}_R(\theta')\cdot x_I(\theta') \; v \; v_R \tag{4}$$

where c_R is supposed to be produced at the time and place of emission and to appear at every point on some continuous trajectory to the right wing; and similarly for c_L, to the left wing. Operation locality requires

$$P(\hat{a}_L(\theta)\cdot\hat{a}_R(\theta')) = P(\hat{a}_L(\theta))P(\hat{a}_R(\theta')) \tag{5}$$

---

[9] 'Locality, Factorability, and the Bell Inequalities', *Nous*, 19 (1985), 485–6.

Coupling (5) with the normal randomness assumptions that the backgrounds, represented by $u_{L(R)}$, $v_{L(R)}$, are independent of each other and of the operation of the causes in all combinations, it is easy to derive that

$$P(x_L(\theta) \cdot x_R(\theta') / \hat{c}_L(\theta) \cdot \hat{c}_R(\theta') \cdot \hat{m}_L(\theta) \cdot \hat{m}_R(\theta')) =$$
$$P(x_L(\theta) / \hat{c}_L(\theta) \cdot \hat{m}_L(\theta)) P(x_R(\theta') / \hat{c}_L(\theta') \cdot \hat{m}_R(\theta'))$$

This is exactly the requirement for full factorizability necessary for the Bell inequality and for other similar no-hidden-variable proofs. It follows immediately that any common-cause model with propagation and locality of operation will contradict quantum mechanics.

It is apparent from this discussion that factorizability is a consequence of conditioning on the hidden states $c_L(\theta)$ and $c_R(\theta')$; but it is the propagation requirement that guarantees the existence of these states. Is propagation, then, a reasonable demand to put on a causal theory in quantum mechanics? Clearly one should be wary, since the difficulties of providing continuous space–time descriptions are arguably the oldest and most worrying in quantum mechanics. The two-slit experiment is the familiar philosopher's illustration of these difficulties, and it will serve here. If propagation is assumed, exactly the wrong predictions are made in the two-slit experiment.

In this experiment, a beam of particles emitted from a source passes through a wall, which contains two separated slits, and finally falls onto a recording screen which registers positions. Two different kinds of pattern are observed on the recording screen. With only one or the other of the two slits open, the particles tend to bunch around a point on the screen directly in line with the slit. But when both slits are open, something very different occurs—the interference pattern. This is exactly the kind of pattern that one would expect if the beam had been composed, not of particles, each of which must pass through one slit or the other, but rather of a wave, whose successive fronts pass through both slits at once. This suggests the conclusion that the particles do not have well-defined trajectories which pass through either slit exclusively.

What is important for the argument is that there are regions on the recording screen where the probability for a particle to register is high with either one slit or the other open, but is practically zero with both. That will be impossible in a causal structure that satisfies local operation and propagation. I will use the simplest model to highlight the essentials of the argument. Nevertheless, I put the discussion in an appendix (Appendix II) because the details are of no special interest. The derivation in Appendix II is similar to the usual derivation that shows that the particles must be treated as waves at the slits. It differs in two respects: first, the quantum particles themselves are allowed to spread out like waves. It is only the causal influence that must follow a trajectory. Second, even when there is only one particle in the system at a time, the model allows that more than one chain of influences may be created in the source, and the trajectories of different influences may pass simultaneously through both slits.

The argument in Appendix II shows that there can be no interference pattern on the screen if the propagation condition is satisifed. Any point on the screen must register at least as many counts with both slits open as with one. The model on which its conclusion is based is supplied with only the most primitive causal structure. Just two substantive causal claims are involved: first, that the production of particles at the source is partially responsible for their appearance at the screen; second, that the opening or closing of a distant slit has no causal influence on what happens simultaneously at the other slit. But even this primitive causal structure must be rejected if propagation is required.

6.8. Conclusion

Grant, for the sake of argument, that the realism issue has been settled by the Bell derivation: the separated systems do not have well-defined spins when they reach the measuring apparatuses. What causes the correlated outcomes, then? The natural answer would be, 'A common-cause in the source'. This chapter has argued that there is nothing wrong with this straightforward account. It is not even necessary to look for a hidden variable: the quantum state itself will serve as the joint cause. It is sometimes reported that the Bell derivation shows that the quantum probabilities are incompatible with any causal structure for the EPR set-up. But the models of section 6.5 show that this is a mistake. The probabilities are entirely consistent with a common-cause model appropriate to the situation.

Yet the models do have a peculiar feature. Although they get the probabilities right, they do not permit propagation between the cause and its effects. The common-cause, which exists in the combined systems at the time of interaction, must operate across a temporal gap. This is especially noticeable in experiments like those of A. Aspect, J. Dalibard, and G. Roger,[10] where the orientation of the measuring apparatus is not even set at the time the systems leave the source. In this case the common-cause is only a partial cause; it must combine with features that later obtain in the measuring devices to produce the measurement outcomes. But if the common-cause is to work in this way, it must operate at the time the particles leave the source, say t_0, to produce an effect at some later time, t', and with nothing to carry the causal influence between the two during the period in between. This is indeed a peculiar kind of causality, and probably most of us will find it just as objectionable as we would find a direct influence from one wing to the other. What distinguishes the two accounts?

Throughout this chapter, as throughout the book, my emphasis has been on the probabilities that follow from various specific structures. A crucial feature is built into all the structures: causes must precede their effects. It is for just this reason that outcomes in the two wings of the EPR experiment are not supposed to be able to influence one another directly. The relevant happenings in the separated wings are meant to occur simultaneously with each other; and much experimental effort must go into making that as true as possible. Of course, simultaneity is a relative concept. The proper past is past in any frame of reference, and it is in this sense that causes are meant to precede their effects. Sometimes one says that events in the proper past are events which are light-ray connectable; and this may somehow suggest that causes must *be* connected with their effects by something like a propagating light-ray. But this kind of propagation is well known not to work in quantum mechanics. It is true that the lessons of the two-slit experiment for quantum realism have not been altogether clear, and there have been a variety of proposals to save particle trajectories. But, in general, even theories which do provide the particles with a continuous path from the source to the recording screen do not propagate the causal influence along a trajectory as well. David Bohm's early hidden-variable models and the traditional pilot wave theory are good

[10] 'Experimental Tests of Bell's Inequalities Using Time Varying Analyzers', *Physical Review Letters*, 49 (1982), 1804–7.

examples. In both cases the quantum state is a principal influence in determining where the particles will be, and the quantum state does not propagate along a trajectory. Whether or not the two-slit experiment refutes realism, it seems to rule out causal propagation, and that is the point of the simple model in Appendix II.

What, then, of the bearing of the Bell inequalities on causality? It seems there are two attitudes one can take. EPR admits a familiar causal structure with proper time-order. It does not admit propagation of influence from cause to effect. Given the familiar lessons of the two-slit experiment, one may think that propagation is a crazy demand to put on a causal structure in quantum mechanics. In that case, the recent work of Bell will be irrelevant to questions of causality in quantum mechanics. Or one may wish for something firmer than the two-slit argument against propagation. The Bell inequalities provide that. But in any case it should be remembered that the connection between propagation and the inequalities is through factorizability; and although factorizability is a consequence of propagation in its most simple and natural form, more recondite assumptions may admit propagation without factorizability, just as the more elaborate models of Arthur Fine admit realism without factorizability. What follows most certainly from the arguments here is that, whatever the case with propagation, factorizability must not enter the argument as a common-cause condition.

Appendix I

A More General Common-Cause Model for EPR

The aim is to construct a model for the EPR arrangement that explicitly includes the apparatuses. The critical assumption in considering the apparatuses and their states is that the two measurements are supposed to be made simultaneously. Thus the state of the apparatus in one wing at the time of measurement can have no effect on the outcome in the other. Indeed, in the experiments of Aspect *et al.*[1] the choice of direction is made at the very last instant, to ensure that no causal connection is possible. Some other assumptions will make the notation easier. Suppose for simplicity that in any complete cause in which states of the apparatus figure as a part, it is always the same apparatus states. Designate the set of states positively relevant to spin-up outcomes along direction θ in the left-wing apparatus, $m_{L,\theta}$, those for the apparatus in the right wing, $m_{R,\theta}$. Allowing initially all combinations of possible partial causes, set

$$x_L(\theta) \equiv V_h V_i V_j V_k \, \hat{a}_{L,\theta}(h,\,i,\,j,\,k) \cdot x_1{}^h \cdot \hat{m}^i_{L,\theta} \cdot \hat{m}^j_{R,\theta'} \cdot x_R(\theta')^k \; v \; u_L$$
$$x_R(\theta') \equiv V_h V_i V_j V_k \, \hat{a}_{R,\theta'}(h,\,i,\,j,\,k) \cdot x_1{}^h \cdot \hat{m}^i_{L,\theta} \cdot \hat{m}^j_{R,\theta'} \cdot x_L(\theta)^k \; v \; u_R$$

where $h,\,i,\,j,\,k \;\epsilon\{0,\,1\}$.

What kinds of causal connection are appropriate for an EPR experiment? The usual assumptions are

1. The state of the measuring apparatus on the right (left) does not influence the outcome on the left (right). This is expressed formally in the model by setting

$$\hat{a}_{L,\theta}(h,\,i,\,1,\,k) = 0 = \hat{a}_{R,\theta'}(h,\,1,\,j,\,k)$$

2. The outcome on the right (left) does not affect the outcome on the left (right). So

$$\hat{a}_{L,\theta}(h,\,i,\,j,\,1) = 0 = \hat{a}_{R,\theta'}(h,\,i,\,j,\,1)$$

3. x_1 is not a complete cause of either the outcome on the left nor of the one on the right:

$$\hat{a}_{L,\theta}(1,\,0,\,0,\,0) = 0 = \hat{a}_{R,\theta'}(1,\,0,\,0,\,0)$$

[1] A. Aspect, J. Dalibard, and G. Roger, 'Experimental Tests of Bell's Inequalities Using Time Varying Analyzers', *Physical Review Letters*, 49 (1982), 1804–7.

4. The state of the apparatus on the left (right) is not a complete cause of the outcome on the left (right):

$$\hat{a}_{L,\theta}(0, 1, 0, 0) = 0 = \hat{a}_{R,\theta'}(0, 0, 1, 0)$$

5. x_1 and $m_{L,\theta}$ are joint partial causes of $x_{L,\theta}$; and x_1 and $m_{R,\theta'}$, of $x_{R,\theta'}$:

$$\hat{a}_{L,\theta}(1, 1, 0, 0) = 0 = \hat{a}_{R,\theta'}(1, 0, 1, 0)$$

Relabelling $\hat{a}_{L,\theta}$ (1, 1, 0, 0) as $\hat{a}_{L,\theta}$; and $\hat{a}_{R,\theta'}$ (1, 0, 1, 0) as $\hat{a}_{R,\theta'}$, the basic equations of the model will be:

$$x_L(\theta) \equiv \hat{a}_{L,\theta} \cdot x_1 \cdot \hat{m}_{L,\theta} v \, u_L$$
$$x_R(\theta') \equiv \hat{a}_{R,\theta'} \cdot x_1 \cdot \hat{m}_{R,\theta'} v \, u_R$$

Appendix II

Do Quantum Causes Propagate?

The arrangement for a two-slit experiment is described in section 6.7. There it was claimed that the quantum-mechanically predicted results show that the propagation requirement is not satisfied in the two-slit experiment. To see why, consider a region on the recording screen with low probability in the two-slit experiment, but high probability with only one slit open—say, slit one. Assume that some feature c operating at t_0 in the source is partially responsible for a positive result, r, at t_2 in the selected region of the screen. By the requirement for propagation, there must be some c_r operating in the region of r, just before t_2, which has been caused by c and which is part of the cause of r. As with EPR, suppose that the measuring apparatus state, m_r, also contributes. Then

$$r(t_2) \equiv a_r(t_2 - \Delta t) \cdot \hat{c}_r(t_2 - \Delta t) \cdot \hat{m}_r(t_2 - \Delta t) \vee u_r \tag{1}$$

The causal influence c_r must propagate to r from c, and since it can appear only where the quantum state is non-zero, it—or some causal antecedent to it—must appear either at slit one at t_1 when the beam passes the wall, or at slit two. Call the causal antecedent at slit one, c_1, and at slit two, c_2. Let s_1 denote the state of the first slit when it is open at t_1; s_2, the second. Allowing that s_1 and s_2 may themselves affect the propagating influence, and assuming for simplicity that there are no further internal contributions, gives the following structure:

$$\hat{c}_r(t_2 - \Delta t) \equiv \hat{a}_1''(t_1 + \Delta t) \cdot \hat{c}_1'(t_1 + \Delta t) \vee \hat{a}_2''(t_1 + \Delta t) \cdot \hat{c}_2'(t_1 + \Delta t') \vee v_r \tag{2}$$

$$\hat{c}_1'(t_1 + \Delta t) \equiv \hat{a}_1'(t_1) \cdot \hat{c}_1(t_1) \cdot \hat{s}_1(t_1) \vee w_1$$

$$\hat{c}_2'(t_1 + \Delta t) \equiv \hat{a}_2'(t_1) \cdot \hat{c}_2(t_1) \cdot \hat{s}_2(t_1) \vee w_2 \tag{3}$$

$$\hat{c}_1(t_1) \equiv \hat{a}_1(t_0) \cdot \hat{c}(t_0) \vee z_1$$

$$\hat{c}_2(t_1) \equiv \hat{a}_2(t_0) \cdot \hat{c}(t_0) \vee z_2 \tag{4}$$

$$\hat{s}_1(t_1) \equiv u_1$$

$$\hat{s}_2(t_1) \equiv u_2 \tag{5}$$

$$\hat{c}(t_0) \equiv u_0 \tag{6}$$

$$\hat{m}_r(t_2 - \Delta t) \equiv v_m \tag{7}$$

Equation (3) contains an important and not yet noted assumption about the causal structure: s_2 appears nowhere in the equation for c_1 nor s_1 in the equation for c_2. This reflects the conventional assumption that the opening or closing of a distant slit has no effect at the other. If it were to do so, the general constraint on structural models that causes must precede their effects would require the distant opening or closing to occur earlier than its effect, and the demand for propagation requires that some capacity-bearing signal should move along a continuous path between.

What is the probability for r to occur, with both slits open? Dropping the time indices, and omitting all terms in which the errors figure:

$$P(\hat{c}_r/\hat{s}_1 \cdot \hat{s}_2) \approx P(\hat{a}_1''\hat{c}_1'/\hat{s}_1 \cdot \hat{s}_2) + P(\hat{a}_2''/\hat{s}_1 \cdot \hat{s}_2) - P(\hat{a}_1'' \cdot \hat{a}_2'' \cdot \hat{c}_1' \cdot \hat{c}_2'/\hat{s}_1 \cdot \hat{s}_2) \geqslant larger$$
$$of \{P(\hat{a}_1'' \cdot \hat{c}_1'/\hat{s}_1 \cdot \hat{s}_2), \ P(\hat{a}_2'' \cdot \hat{c}_2'/\hat{s}_1 \cdot \hat{s}_2)\}$$

Let the larger be $P(\hat{a}_1'' \cdot \hat{c}_1'/\hat{s}_1 \cdot \hat{s}_2)$. With the usual assumptions that the operations are independent of the causes and the errors of each other

$$P(\hat{a}_1'' \cdot \hat{c}_1'/\hat{s}_1 \cdot \hat{s}_2) \approx P(\hat{a}_1'' \cdot \hat{a}_1' \cdot \hat{a}_1)P(u_0/u_1 \cdot u_2) = P(\hat{a}_1'' \cdot \hat{a}_1' \cdot \hat{a}_1)P(u_0/u_1 \cdot \neg u_2) =$$
$$P(\hat{a}_1'' \cdot \hat{c}_1'/\hat{s}_1 \cdot \neg \hat{s}_2) \approx P(\hat{c}_r/\hat{s}_1 \cdot \neg \hat{s}_2)$$

It follows that

$$P(\hat{r}/\hat{s}_1 \cdot \hat{s}_2) \approx P(\hat{a}_r)P(\hat{m}_r)P(\hat{c}_r/\hat{s}_1 \cdot \hat{s}_2) \geqslant P(\hat{a}_r)P(\hat{m}_r)P(\hat{c}_r/\hat{s}_1 \cdot \neg \hat{s}_2) \approx P(\hat{r}/\hat{s}_1 \cdot \neg \hat{s}_2)$$

Hence the probability for a particle to register at r with both slits open is at least as large as the larger of the probabilities with only one slit open. This is contrary to what happens.

Appendix III

Propagation, Effect Locality, and Completeness: A Comparison

In discussing the Bell inequalities, most authors do not separate questions of realism from questions of causality. Among the few who do address questions of causality separately, propagation is usually presupposed. Patrick Suppes is a good example. He produces some nice results on exchangeability that suggest that the requirement of conditional independence of the outcomes on the common-cause—i.e. factorizability—should be given up: 'The demand for conditional independence is too strong a causal demand.' The considerations in this chapter show why it is too strong. As 'a way of finding a new line of retreat', Suppes suggests making instead an 'independence of path assumption'. This assumption 'prohibits of course instantaneous action at a distance and depends upon assuming that the propagation of any action cannot be faster than that of the velocity of light'.[1] In its most straightforward form, Suppes's suggestion amounts to the requirement for propagation.

A requirement for propagation also appears in the work of Jon Jarrett.[2] Jarrett derives 'full factorizability' (recall section 6.5) from two simpler conditions. *Completeness* requires that the outcomes not cause each other; they are to depend on some prior state of the combined systems—say x_1—and the states of the measuring apparatus. Jarrett puts the condition probabilistically. In the notation used here it reads thus:

Completeness (Jarrett)

$$P(x_L(\theta) \cdot x_r(\theta') / x_1 \cdot \hat{m}_L(\theta) \cdot \hat{m}_R(\theta')) =$$
$$P(x_L(\theta) / x_1 \cdot \hat{m}_L(\theta) \cdot \hat{m}_R(\theta')) P(x_R(\theta') / x_1 \cdot \hat{m}_L(\theta) \cdot \hat{m}_R(\theta'))$$

Jarrett names his other condition *locality*. It is exactly the same as the first assumption of common-cause model No. 2 in section 6.5: the state of the measuring apparatus in one wing must have no effect on the outcome in the other. Again, Jarrett does not put the condition directly, but rather in terms of its probabilistic consequences:

---

[1] P. Suppes, 'Causal Analysis of Hidden Variables', in P. Asquith and R. Gieve (eds.), *PSA [proceedings of the biannual Philosophy of Science Association meetings] 1980*, ii, (East Lansing, Mich.: Philosophy of Science Association, 1980), 529.

[2] 'On the Physical Significance of the Locality Condition in the Bell Arguments', *Nous*, 18 (1984), 569–89.

Locality (Jarrett)

$$P(x_L(\theta)/x_1 \cdot \hat{m}_L(\theta) \cdot \hat{m}_R(\theta')) = P(x_L(\theta)/x_1 \cdot \hat{m}_L(\theta))$$
$$P(x_R(\theta')/x_1 \cdot \hat{m}_L(\theta) \cdot \hat{m}_R(\theta')) = P(x_R(\theta')/x_1 \cdot \hat{m}_R(\theta'))$$

These, too, are factorization conditions, although superficially they may not seem so.[3]

The locality assumption is familiar from the discussion of full factorizability in section 6.5. It rests on the supposition that the two measurements are made simultaneously, and it may thus conceal a commitment to propagation. That depends on whether the distant apparatus is ruled out as a cause on the grounds of time-order alone—causes must precede their effects; or on the more complex grounds that the relativistic requirement that no signal can travel faster than the speed of light rules out any possibility of propagation (at the time of measurement) from the apparatus in one wing to the outcome in the other. But whatever is the case with Jarrett's locality assumption, his completeness requirement puts him squarely in favour of propagation.

In defending the completeness requirement, Jarrett sometimes talks about 'screening off'. This is language taken from discussions of the common-cause. But we have seen that this kind of factorizability condition is not an appropriate one for common causes in EPR. Nor does it really seem to be what Jarrett has in mind. For he does not in fact treat the hidden states as descriptions of the common cause at the source, but rather as features that the two systems possess separately at points between the source and the measurement. This is apparent in the example he gives to explain why he chooses the name 'completeness'. In the example, each of the two systems is supposed to have a genuine spin in some particular direction after it leaves the source, and that spin deterministically causes the measurement outcome in every other direction. Jarrett points out that in this case the measurement outcomes do not factor when information about the actual spins is omitted. Factorizability is restored when the description of the systems at the time of measurement is complete. This he claims to be generally true:

For a theory which assigns probabilities on the basis of the complete state description . . . any information about particle *L* (*R*) [*the particle in the left wing (the particle in the right wing)*] which may be inferred from the outcome of a measurement on particle *R* (*L*) [*the particle in the right wing (the particle in the left wing)*] is clearly redundant.[4]

---

[3] They are equivalent to
$$P(x_L(\theta) \cdot \hat{m}_R(\theta')/x_1 \cdot \hat{m}_L(\theta)) = P(x_L(\theta)/x_1 \cdot \hat{m}_L(\theta)) P(\hat{m}_R(\theta')/x_1 \cdot \hat{m}_L(\theta))$$
and
$$P(x_R(\theta') \cdot \hat{m}_L(\theta)/x_1 \cdot \hat{m}_R(\theta')) = P(x_R(\theta')/x_1 \cdot \hat{m}_R(\theta')) P(\hat{m}_L(\theta)/x_1 \cdot \hat{m}_R(\theta'))$$

[4] Jarrett, op. cit., p. 580.

Clearly the criticism only works for theories—like the one Jarrett proposes—which postulate that some intervening states exist. A theory can only be accused of incompleteness if there is something with respect to which it could be complete. So in calling his condition *completeness*, Jarrett already supposes that there are further states to be described, and that is the matter at issue. The completeness assumption presupposes both that there are spatially separated and numerically distinct states in each of the two systems between the time of interaction and the time of measurement, and also that these states contain all the information from the past histories of the systems which is causally relevant to the measurement results. Where do such states come from? They are no part of the common-cause models; but they are just what is needed for propagation.

Linda Wessels is more explicit in her assumptions than Jarrett. Her paper 'Locality, Factorability, and the Bell Inequalities'[5] provides a long treatment of a concept she calls 'effect locality'. This paper is a unique contribution in the Bell literature, important because of the way it tries to get behind the factorizability condition to find out where it comes from. Wessels claims that the condition is a consequence of the demand that there is no action at a distance. This means that factorizability will turn out to be, for Wessels, a kind of propagation requirement.

Wessels describes effect locality 'very roughly and intuitively', thus:

the evolution of the characteristics associated with a body B 'in a moment', say (t,t + dt), depends *only* on the characteristics associated with B at t and the external influences felt by B during the moment. A slightly different but equally rough version of effect locality is the following two part claim: a) the past experience of B influences its own evolution at t only in so far as that past experience has either led to the characteristics associated wtih B at t or/and has influenced other systems (bodies, fields, whatever) which in turn exert external influences on B at t; and b) other systems influence the evolution of B at t only to the extent that they are the source of some external influence (force, field, whatever) acting on B in (t,t + dt)[6].

The bulk of Wessels' paper is devoted to presenting a more detailed and precise characterization from which a factorizability condition can be derived.

I will take up three topics in discussing Wessels' idea. The first will be a brief characterization of some of the features of her formalism that are missing from the more simplified accounts I have been presenting; the second will show why effect locality and propagation are the same; and the third will return to the basic naive assumptions with which worries about EPR correlations begin.

[5] *Nous*, 19 (1985), 481–519.
[6] Ibid. 489–90.

(*a*) Wessels introduces a number of sophistications which I have omitted because they are not essential to the central ideas here. I have tried throughout to make as many simplifications as possible in order to keep the fundamental outlines clear. But there are a number of features present in Wessels' formalism that can be taken over to make my models more general. For a start, Wessels separates ' "influences" by which systems can affect one another . . . [which] might be characterized as forces, potentials, field strengths, or any other "felt influence" '[7] from the conditions and states of other bodies which might affect the state of a body contiguous to them. (But in every case she requires that 'the quantitative properties involved are all empirically measurable': 'The morning tide is not explained merely by saying vaguely, "Ah, there is a force acting at a distance." A distance force has a quantitative measure.'[8]) In the kinds of simple structural equation considered here, this distinction is not drawn: all measurable quantities are represented in the same way by random variables.

Second, Wessels allows a variety of intervening states which can have different degrees of effect, whereas only the simplest case, where the propagating influence is of a single kind, is taken up here. Thirdly, she provides explicitly for the time evolution of the capacity-carrying states, and she allows for interactions along the path between the initial cause and the final effect (cf. her clause (E L4) and (E L2) respectively), something which is missing both from my account of propagation and from the at-at theory of Wesley Salmon on which it is modelled. Lastly, Wessels uses continuous densities and continuous time, which is a far more satisfactory way to treat questions of contiguity. In all these respects, Wessels provides a framework within which a more general formulation could be given of the basic models described in the earlier sections of this chapter. But I shall not do that here.

(*b*) A number of the conditions that define an effect-local theory serve to ensure the existence and law-like evolution of whatever states the systems may have. These conditions will be discussed below. Only three concern the causal connections among the states; and all three are formulated probabilistically by Wessels. The first is (EL3): 'EL3 says that the evolution of a body B over a period of time . . . depends only on the state of B at the beginning of the period and the influences at the positions of B during that period.'[9] Specifically excluded are earlier states of the body, and any other circumstances that might surround it. Letting $\{x'_h\}$ represent the set of states available at the end of the interval, $\{x_i\}$ the set of positively relevant states at the beginning, and $\{F_j\}$ the influences at the positions of B during the interval, the structural form for (EL3) is

[7] Ibid. 490.
[8] Ibid. 484.
[9] Ibid. 491.

$$x'_h \equiv V_i V_j \hat{a}_{hij} \cdot x_i \cdot \hat{F}_j \, v \, u_h$$

Wessels does not put the condition structurally, but instead expresses it directly in terms of the probabilities. Representing the other circumstances by C and the set of earlier states by $B(t)$, her probabilistic formulation, in the notation used here, reads thus:

$$P(x'_h / x_i \cdot \hat{F}_j \cdot \hat{B}(t) \cdot \hat{C}) = P(x'_h / x_i \cdot \hat{F}_j) \qquad \text{(EL3′)}$$

What matters for the EPR case is that the excluded circumstances 'include states of bodies other than B'.[10] The set of states available to the distant system in the EPR experiment will be represented by $\{y_k\}$. Referring to these states explicitly, (EL3) implies the familiar factorization condition

$$P(x'_h / x_i \cdot \hat{F}_j \cdot y_k) = P(x'_h / x_i \cdot \hat{F}_j)$$

The second causal restriction in an effect-local theory is (EL6)

Effect locality requires that the way an interaction proceeds depends on nothing more than the states of the interacting systems just before the interaction (then regarded as bodies) and the influences external to the interaction at the location of and during the course of interaction. Other earlier states of the interacting systems, or of other systems, as well as influences evaluated at other places and/or at earlier times, are statistically irrelevant. This requirement is captured by Clause EL6.[11]

Letting E_{12} represent the external influences on the interacting bodies, $B_1(t)$ and $B_2(t)$, as before, the set of prior states of the two bodies, and C_{12} the circumstances:

$$P(x'_h \cdot y'_k / x_i \cdot y_j \cdot \hat{E}_{12} \cdot \hat{B}_1(t) \cdot \hat{B}_2(t) \cdot \hat{C}_{12}) = P(x'_h \cdot y'_k / x_i \cdot y_j \cdot \hat{E}_{12}) \qquad \text{(EL6)}$$

Again, this is a familiar-looking factorization condition.

Thirdly, 'something like an assumption of sufficient cause for correlation is needed:

SC. If there has been nothing prior to or at a time t that causes a correlation of the states of two independent systems, then their states are uncorrelated at t.'[12]

In the application to EPR, (SC) is joined with the assumption usually made about the measuring apparatuses, that they have no causal history in common, to produce factorization of the outcomes on the apparatus states. This is the assumption that makes the theory 'local' in Jarrett's terminology. The real significance of Wessels' work comes from her characterization of

[10] Ibid.
[11] Ibid. 493.
[12] Ibid. 508.

effect locality, which provides the grounds for the condition that Jarrett calls 'completeness'.

Wessels gives the condition (SC) a new kind of name because she thinks that it is not part of the characterization of an effect-local theory. She says, 'SC is certainly in the spirit of effect-locality, but it does not follow from effect-locality as explicated above. It must be taken as an independent assumption.'[13] On the contrary, I think there is more unity to be found in Wessels' conditions than she claims; for something like her assumption that a sufficient cause is required for correlation is already presupposed in her statistical characterizations (EL3) and (EL7). These are assumptions about causality: (EL3) is meant to assert that the state of the system in one wing is not causally relevant to the outcome in the other; and (EL7), that the only causes for the outcomes of an interaction between two bodies are the ingoing states of the bodies and the influences contiguous to them. Yet in both cases the formulation that Wessels gives to the conditions is probabilistic. (EL3) is supposed to say that F and x are the only causes of x'; earlier states play no causal role, nor do any of the states in the distant system. But what ensures that these states are statistically irrelevant? Something like Reichenbach's Principle of the common-cause is needed to bridge the gap; and (SC) is Wessels' version of it. This just confirms the point from earlier chapters that you cannot use probabilities to measure causality unless you build in some appropriate connection between causality and probability at the start.

A more substantial concern about the characterization of effect locality concerns factorizability. (EL3) is, after all, just the factorization condition that I have argued is unacceptable as a criterion in EPR. So when $F_j.x_i$ is a common cause of both $x_{h'}$ and $y_{k'}$, (EL3') will not generally be valid. But this is not a real problem in Wessels' derivation, because the use to which she puts (EL3) is more restricted than her general statement of it. In fact (EL3) is not used at any point where the antecedent state in question might be a common-cause. This is apparent from a brief review of Wessels' strategy, even without inspection of the details of her proof.

Wessels aims to produce an expression for the joint probability of the final outcome states in an EPR-type experiment. To do so, she starts at the end and works backwards. The probability for the outcomes is calculated by conditionalization from the probabilities just before, and those in turn by conditionalization on the states just before them, and so on in a chain backwards, each step employing (EL3). It is only at the first step, at the point of interaction, that (EL3) might fail, and that makes no difference; the last step is what matters. It is apparent both from the discussion of propagation in section 6.7 and from the discussion of Jarrett in this section, that factorizability is a trivial consequence once the individual systems have their own distinct states which are causally responsible for what happens to

[13] Ibid.

them thereafter. Indeed, Wessels' scheme satisfies all the requirements for propagation, since the travelling particles are each assigned states, localized along their trajectories, from the instant the interaction between them ceases.

This raises again the question: why assume that such states exist? Jarrett, we have seen, takes their existence as obvious. The propagation requirement of section 6.7 says that they are necessary if features of the interaction are going to make any difference to the outcomes. Wessels gives a different reason, and this is the point of most of the remaining conditions on effect-local theories; namely, once the interaction has ceased, the two separated systems should be treated as bodies, and bodies, in an effect-local theory, ought to have their own states.

(c) This notion of body is the third topic that I want to take up. I believe that Wessels is right in her characterization, and that it is just this notion of a body with localized states that lies at the core of the common refusal to attribute causal structures in quantum mechanics. This is in no way a novel view. The Einstein–Podolsky–Rosen paper was written partly in response to Niels Bohr's claims about the wholeness of quantum actions, and the paper precipitated Erwin Schroedinger's very clear statement of the entanglement of separated quantum systems.[14]

I will discuss wholeness versus separability of bodies briefly here to reinforce my conclusions about EPR and causality. There is a tendency to think that the distant correlations of an EPR-type experiment provide special obstacles to the postulation of causal structures in quantum mechanics. The general point of this chapter is that this is not so. The arrangement of probabilities is not a particular problem. What is a problem is the localizability of causal influence. Recall the discussion of the two-slit experiment in Appendix II. The derivation there of a wrong prediction for the pattern on the screen looks much like the standard derivation of that result. But it is not quite the same; for what is critical in Appendix II is not where the particle is, but where the causal influence is. The derivation requires that the influence be highly localized; this is the requirement that comes in Jarrett under the name 'completeness', in section 6.7 from the demand for propagation, and in Wessels from the treatment of the systems as bodies. If, by contrast, the causal influence is allowed to stretch across both slits—as the quantum state does—no problem arises. Under its old-fashioned name, this is the problem of wave-particle duality. The point I want to make here is that it is the very familiar problem of wave-particle duality that prevents a conventional causal story in EPR, and not its peculiar correlations.

---

[14] This was the paper which introduced his famous cat paradox. Cf. A. Fine, *Shakey Game* (Chicago, Ill.: Chicago University Press, 1987).

One quick way to see this is to consider an alternative to the quantum structure proposed in section 6.6. There the suggestion is to take the quantum state at the time of interaction as the common cause of the two measurement outcomes. The succeeding section shows that the influence from this cause cannot propagate along a trajectory. I think of it instead as operating across a gap in time. An alternative is to let it propagate, like the quantum state, as a 'wave'—or, finally, just to take the quantum state itself at the time of measurement as the immediate cause of both outcomes. That is really where the discussion of the EPR correlations starts, before any technicalities intervene. What is the idea of causation that prevents this picture? It is a good idea to rehearse the basic problem.

No one can quarrel with the assumption that a single cause existing throughout a region r can produce correlated effects in regions contiguous to r. That is a perfectly classical probabilistic process, illustrated by the example in section 6.3 and by a number of similar ones constructed by van Fraassen in his debate with Salmon. An extremely simplified version comes from Patrick Suppes. A coin is flipped onto a table. The probability for the coin to land head up is ½. But the probability for head up and tail down $\neq$ ½ × ½. The structure of the coin ensures that the two outcomes occur in perfect correlation. Similarly, if the quantum state for the composite at $t_2 - \Delta t$ is allowed as a cause, then the EPR set-up has a conventional time-ordered common-cause structure with both propagation and a kind of effect locality. There is a state immediately preceding and spatially contiguous with the paired outcomes; the state is causally relevant to the outcomes; and this state, after the addition of whatever contiguous felt influences there are, is all that is relevant. But it does not produce factorization, and hence does not give rise to the Bell inequalities. What rules out this state as a cause?

Wessels rules it out by maintaining that the two separated systems are distinct bodies, with spatially separated surfaces. Her justification for this is her requirement (EL4): an effect-local theory will treat any system as a distinct body so long as it is not interacting with any other. It is the extent of the surface of the body that is the problem. There are three apparent choices: (1) something approximating to the classical diameter for the electron; (2) the spatial extent of the left-travelling half of the wave packet for the body in the left wing, and of the right-hand travelling half for the body in the right wing; or (3) the total extent of the wave packet. The first choice gives factorizability, but this is just the assumption that goes wrong in the two-slit experiment. The third is a choice that will not present the outcome probabilities as factors. The second is the one she needs, and that choice is hard to defend. It would be reasonable if one were willing to assume that the wave packet reduces when the two systems stop interacting; but that assumption will give the wrong predictions. Without reduction of the wave packet it is a strange blend, both taking quantum mechanics seriously and refusing to do so: the electrons are allowed to spread out in the peculiar way that quantum systems do, far enough to pass through both slits at once, for example; but the two electrons are kept firmly in the separate spread-out halves.

Imagine a single particle that strikes a half-silvered mirror. After a while the quantum state will have two almost non-overlapping humps, where each hump itself will be much wider than the classical dimensions of the particle, due to quantum dispersion. Where is the particle? Here is the problem of wave-particle duality. The 'surface' of the particle must spread like a wave across both humps. Quantum mechanics cannot locate it in either one hump or the other. In the spatial portion of the wave packet in E P R there are again two humps, although this time there are two particles. But nothing in quantum mechanics says that one particle is in one hump and the other in the other. If it is formulated correctly to express the assumptions of the problem, the theory will say that a measurement of charge or mass will give one positive answer on the right and one on the left, and never two on the same side; just as, with the half-silvered mirror, the theory should predict a positive reading on one side of the mirror or the other, but not on both sides at once. But that does not mean in either case that a particle can be localized in either of the two humps exclusively. There is good intuitive appeal in keeping particles inside their classical dimensions. But once they are spread out, what is the sense in trying to keep one very smeared particle on the left and the other on the right? It is just that assumption that makes E P R seem more problematic, causally, than older illustrations of wave-particle duality.

Index